Multimodal Video Characterization And Summarization

THE KLUWER INTERNATIONAL SERIES IN VIDEO COMPUTING

Series Editor

Mubarak Shah, Ph.D.
University of Central Florida
Orlando, USA

Other books in the series:

3D FACE PROCESSING: MODELING, ANALYSIS AND SYNTHESIS
Zhen Wen, Thomas S. Huang ; ISBN: 1-4020-8047-6

EXPLORATION OF VISUAL DATA
Xiang Sean Zhou, Yong Rui, Thomas S. Huang ; ISBN: 1-4020-7569-3

VIDEO MINING
Edited by Azriel Rosenfeld, David Doermann, Daniel DeMenthon;ISBN: 1-4020-7549-9

VIDEO REGISTRATION
Edited by Mubarah Shah, Rakesh Kumar; ISBN: 1-4020-7460-3

MEDIA COMPUTING: COMPUTATIONAL MEDIA AESTHETICS
Chitra Dorai and Svetha Venkatesh; ISBN: 1-4020-7102-7

ANALYZING VIDEO SEQUENCES OF MULTIPLE HUMANS: Tracking, Posture Estimation and Behavior Recognition
Jun Ohya, Akita Utsumi, and Junji Yanato; ISBN: 1-4020-7021-7

VISUAL EVENT DETECTION
Niels Haering and Niels da Vitoria Lobo; ISBN: 0-7923-7436-3

FACE DETECTION AND GESTURE RECOGNITION FOR HUMAN-COMPUTER INTERACTION
Ming-Hsuan Yang and Narendra Ahuja; ISBN: 0-7923-7409-6

Multimodal Video Characterization And Summarization

Michael A. Smith
AVA Media Systems
Carnegie Mellon University

Takeo Kanade
Carnegie Mellon University

KLUWER ACADEMIC PUBLISHERS

Dr. Michael A. Smith
France Telecom Research & Development
801 Gateway Blvd, Suite 500
South San Francisco, CA 94080
U.S.A.

Email: msmith@savasystems.com

Dr. Takeo Kanade
The Robotics Institute
Carnegie Mellon University
5000 Forbes Avenue
Pittsburgh, PA 15213
U.S.A.

Email: takeo.kanade@cs.cmu.edu

Library of Congress Cataloging-in-Publication Data

A C.I.P. Catalogue record for this book is available
from the Library of Congress.

Smith, Michael A.
Multimodal Video Characterization and Summarization / Michael A. Smith, Takeo Kanade
p.cm.—(Kluwer International series in video computing)
Includes bibliographical references and index.

ISBN 978-1-4419-5351-3

e-ISBN 978-0-387-23008-5

e-ISBN 0-387- 23008-4 Printed on acid-free paper.

9 8 7 6 5 4 3 2 1

springeronline.com

Table of Contents

Series Foreword

Traditionally, scientific fields have defined boundaries, and scientists work on research problems within those boundaries. However, from time to time those boundaries get shifted or blurred to evolve new fields. For instance, the original goal of computer vision was to understand a single image of a scene, by identifying objects, their structure, and spatial arrangements. This has been referred to as image understanding. Recently, computer vision has gradually been making the transition away from understanding single images to analyzing image sequences, or video understanding. Video understanding deals with understanding of video sequences, e.g., recognition of gestures, activities, facial expressions, etc. The main shift in the classic paradigm has been from the recognition of static objects in the scene to motion-based recognition of actions and events.

Video understanding has overlapping research problems with other fields, therefore blurring the fixed boundaries. Computer graphics, image processing, and video databases have obvious overlap with computer vision. The main goal of computer graphics is to generate and animate realistic looking images, and videos. Researchers in computer graphics are increasingly employing techniques from computer vision to generate the synthetic imagery. A good example of this is image-based rendering and modeling techniques, in which geometry, appearance, and lighting is derived from real images using computer vision techniques. Here the shift is from synthesis to analysis followed by synthesis. Image processing has always overlapped with computer vision because they both inherently work directly with images. One view is to consider image processing as low-level computer vision, which processes images, and video for later analysis by high-level computer vision techniques. Databases have traditionally contained text, and numerical data. However, due to the current availability of video in digital form, more and more databases are containing video as content. Consequently, researchers in databases are increasingly applying computer vision techniques to analyze the video before indexing. This is essentially analysis followed by indexing.

Due to MPEG-4 and MPEG-7 standards, there is a further overlap in research for computer vision, computer graphics, image processing, and databases. In a typical model-based coding for MPEG-4, video is first analyzed to estimate local and global motion then the video is synthesized using the estimated parameters. Based on the difference between the real video and synthesized video, the model parameters are updated and finally coded for transmission. This is essentially analysis followed by synthesis, followed by model update, and followed by coding. Thus, in order to solve research problems in the context of the MPEG-4 codec, researchers from different video computing fields will need to collaborate. Similarly, MPEG-7 is bringing together researchers from databases, and computer vision to specify a standard set of descriptors that can be used to describe various types of multimedia information. Computer vision researchers need to develop techniques to automatically compute those descriptors from video, so that database researchers can use them for indexing. Due to the overlap of these different areas, it is meaningful to treat video computing as one entity, which covers the parts of computer vision, computer graphics, image processing, and databases that are related to video. This international series on Video Computing will provide a forum for the dissemination of innovative research results in video computing, and will bring together a community of researchers, who are interested in several different aspects of video.

Mubarak Shah
University of Central Florida, Orlando

Acknowledgements

I would like to thank my family, Mr. and Mrs. Richard Smith Jr., Veneka, Nyasha, Caymen, and Carmen. The authors wish to thank the supporting members of the Carnegie Mellon University Vision and Autonomous Systems Center, including Toshio Sato, Shin'ichi Sato, Yuichi Nakamura, Henry Rowley, Henry Schneiderman, Tsuhan Chen, and Rawesak Tanawongsuwan. The authors also wish to thank the supporting members from the Carnegie Mellon University Informedia Digital Video Library Project, including Howard Wactlar, Michael Christel, Alex Hauptmann, and Scott Stevens.

A portion of the material in this chapter was developed by AVA Media Systems. The Informedia material is based on work supported by the National Science Foundation (NSF) under Cooperative Agreement No. IRI-9817496. A portion of this research is also supported in part by the advanced Research and Development Activity (ARDA) under contract number MDA908-00-C-0037. CNN and WQED Communications in Pittsburgh, PA supplied video to the Informedia library for sole use in research. Their video contributions as well as video from NASA, the U.S. Geological Survey, and U.S. Bureau of Reclamation are gratefully acknowledged.

Michael A. Smith
Current Affiliation - Director, Digital Content Management
France Telecom R&D, San Francisco, CA.

Chapter 1

Introduction

This concerns the field of video characterization and summarization. The term "characterization" refers to methods to analyze video content or some other medium at various levels. We emphasize video content processing, which encompasses methods in image, audio, and language understanding. The term "summarization" refers to techniques for abstracting video without apparent loss in content.

This book is organized as follows: We first introduce the field of video characterization and its relation to professional production standards and research in video understanding. Second, we describe various techniques for characterizing video content. In particular, we define the video skim as one of the most important video summarization techniques. Finally, we introduce subjective user-studies as an effective means to evaluate video characterization systems.

1.1 Video Characterization

The term media encompasses all modalities: audio, image, language and other forms of digital content. It is pervasive in all aspects of communication and information exchange. Video represents a dynamic form of media that is used in entertainment, surveillance, education, and a host of other applications. Video understanding research focuses on the analysis, description and manipulation of the contents of a video for a specific application.

The techniques for analyzing individual modalities have been studied in a number of applications. A methodology for characterizing video, however, must use a combination of image, audio and language technologies, as video is comprehensive media consisting of these modalities. This requires integrating these techniques for extraction of significant information, such as specific objects, audio keywords, and video structure.

In Figure 1.1, the concept of characterization is illustrated for a video titled "Interpreting Brain Signals". The thumbnails at the top represent four shots from this video. The two-tier, low-level and mid-level features, describe the content according to the level of semantic understanding. Low-level features simply represent statistical content such as color, texture, and audio levels, together with the detection of on-screen text, camera motion, object motion, face detection, and audio classification.

A Mid-level feature attempts to interpret content with higher semantics. Examples of mid-level features used in Figure 1.1 include scene classification, video pans, object classification and movement, and speaker introduction. In chapter 2, we describe other mid-level features that may be derived for video production and archival standards. Chapter 3 presents an overview of several methods for automated feature analysis.

The tiered video characterization will be used throughout the book. In general, the number of tiers is not limited to two. This form of characterization is useful when merging multiple media formats, and new or older techniques for image and audio analysis. For example, consider the higher-level problem of image categorization or clustering. In most cases, image categories are determined by some combination of low-level image analysis, including the detection of average color, dominant color, localized color, image texture, and image structure or specific objects.

Characterization Architecture

Figure 1.2 shows different levels of video analysis and their relationship. Automatic interpretation of video content has been an active area of research under many different names and disciplines. Examples are pattern recognition, object detection, and image processing, whose results serve as

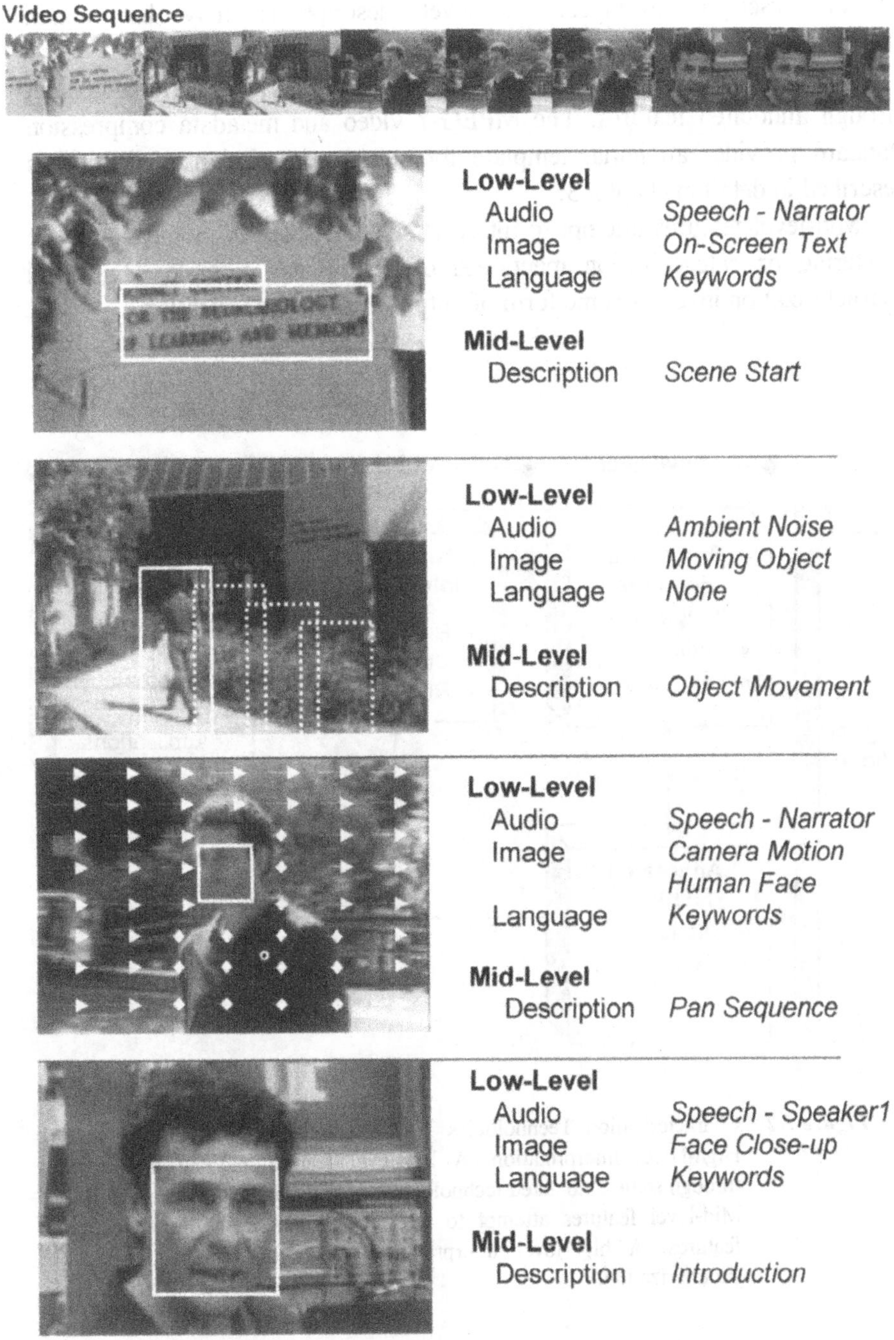

Video Sequence

Low-Level
Audio — *Speech - Narrator*
Image — *On-Screen Text*
Language — *Keywords*

Mid-Level
Description — *Scene Start*

Low-Level
Audio — *Ambient Noise*
Image — *Moving Object*
Language — *None*

Mid-Level
Description — *Object Movement*

Low-Level
Audio — *Speech - Narrator*
Image — *Camera Motion*, *Human Face*
Language — *Keywords*

Mid-Level
Description — *Pan Sequence*

Low-Level
Audio — *Speech - Speaker1*
Image — *Face Close-up*
Language — *Keywords*

Mid-Level
Description — *Introduction*

Figure 1.1 Video Characterization Process Flow. Low-level features come from traditional image, audio and language processing as well as annotated features such as closed captions and other manual data. Mid-level features attempt to interpret video content, such as object motion and on-screen text. High-level features would interpret video through summarization, segmentation, or custom visualization.

low-level descriptions of video. Low-level descriptions may be human annotated.

There has not been a standard format for describing video content through annotated features. The MPEG-7 video and metadata compression standard provides an initial template for content descriptions. MPEG-7 is described in detail in chapter 3.

Mid-level features attempt to interpret semantic content or meaning. The clustering or categorization mentioned earlier is one example. High-level characterization involves some form of output display or application.

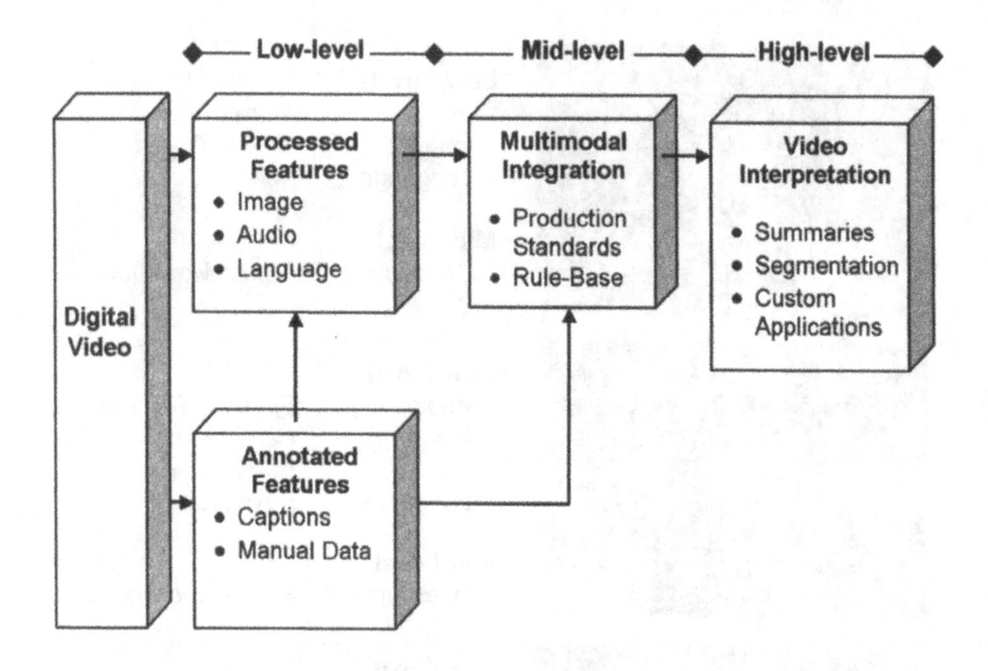

Figure 1.2 Characterization Technologies – Low and Mid-Level Features and High-Level Interpretation. A low-level feature may be processed through some automated technology or provided by manual annotation. Mid-level features attempt to interpret content from the low-level features. A high-level interpretation serves as visualization and summarization.

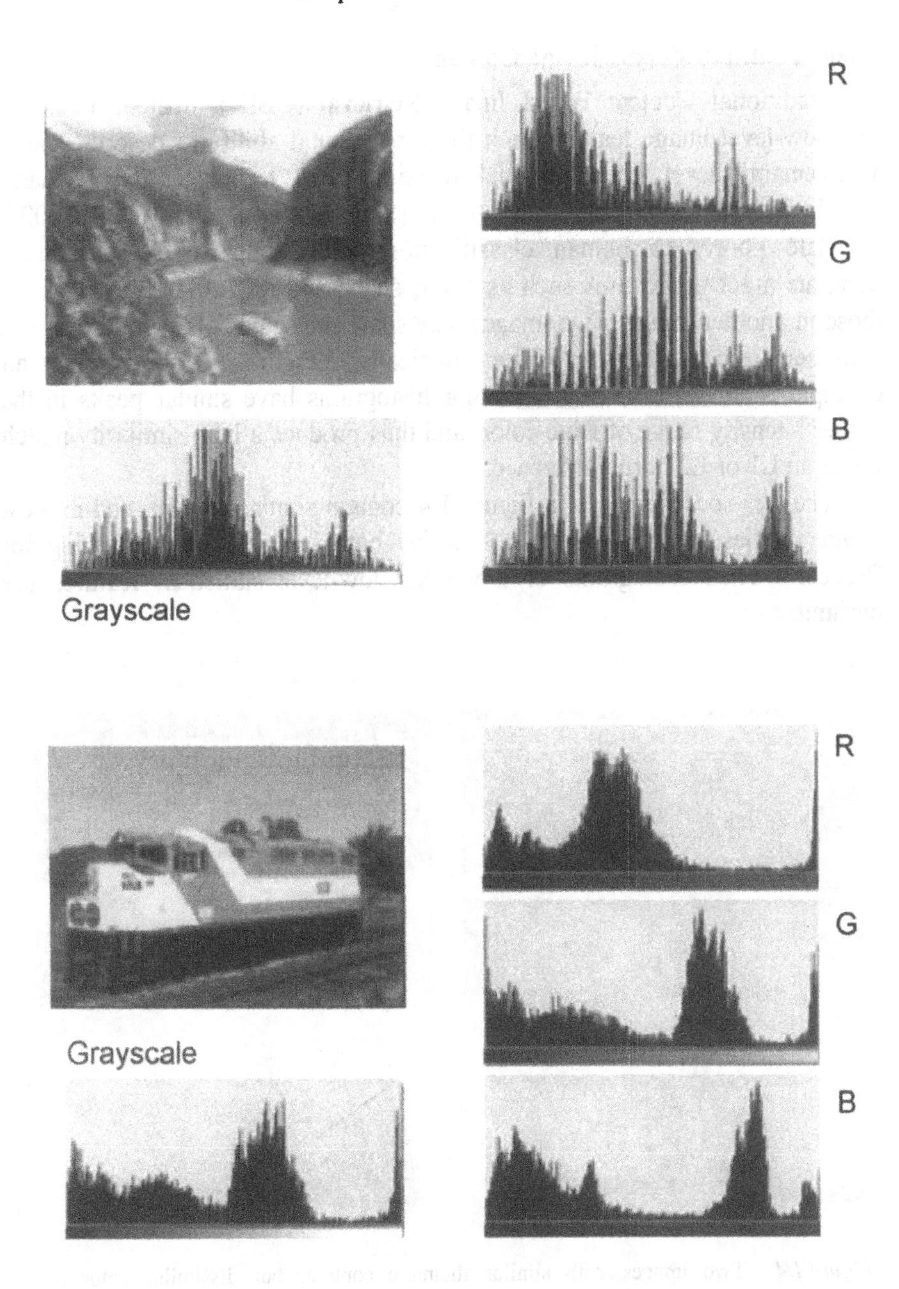

Figure 1.3 Two images with similar RGB color histograms, but with dissimilar thematic content. The bottom image (train) was encoded at a much higher resolution than the top image (gorge). The red and green histograms are different, but the blue histograms exhibit slight peaks at the higher intensity levels. In a comparison with images from a much larger database, the small overlapping peak in the higher intensity blue histograms often results in a matching similarity score between the two images.

Image Features Versus Image Content

Traditional Content Based Image Retrieval (CBIR) methods that use only low-level image features lack the fundamental ability to describe video at a semantic level. The traditional image matching techniques (e.g., [Zhang 1995], [Rubner 1998], [Cohen 1999], [Santini 1999], and [Iqbal 2002]) correlate poorly to human classification and testing. These techniques correlate a set of features such as color, shape and texture in one image, to those in another image. Two images may share the same color properties, but their semantic content can differ drastically. Figure 1.3 shows such an example. The two sets of RGB color histograms have similar peaks in the higher intensity range of Blue color, and thus produce a high similarity match under an L1 or L2 norm difference.

The two soccer images in Figure 1.4 contain similar content, and in most image queries, a user expects to receive both images when searching for "Soccer." Their histograms or any other low-level statistical features are dissimilar.

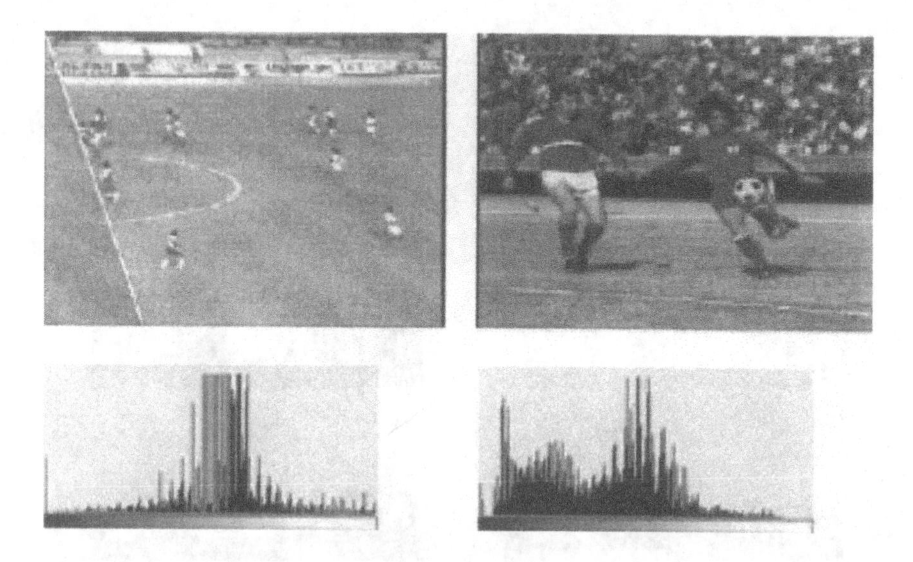

Figure 1.4 Two images with similar thematic content, but dissimilar color or grayscale histograms. The image to the left is grayscale and set as a distance shot. The right image is a close-up shot, with very dissimilar image properties. The grayscale histograms provide little or no insights to the content of the images.

Features for Characterization

Digital video is a composite of image, audio, language and text modalities. Multimodal integration provides video understanding at a level not possible in single modal systems. Combinations of image, audio, language, and annotated features are used to describe content. Semantic video characterization attempts to recognize significant events. These events may be as simple as a shot change or as complicated as a scene change. A shot is a change in visual field of view. A scene is a thematic unit that may consist of events such as a person entering a room, the insertion of some audio effect, or multiple shots.

Low and high-level features are extracted in a hierarchical structure. These features may be manually extracted by some annotation system, or through automated processes. Examples of annotated features include closed captions and video descriptive services (VDS). Automated and annotated features are integrated to interpret video according to predetermined rule-base descriptors. After interpretation, different visualizations of content may occur for different applications. This book uses the video summary as a visualization of semantic video characterization. Figure 1.2 has illustrated the basic architecture for this research.

Video Feature Detection

Low-level feature detection for digital imagery and audio has been studied in detail [Bovik 2000], [Shapiro 1992], [Gonzalez 1992], [Furht 1996], and [Furht 1999]. This vast body of research contains well-known techniques for things such as shot change detection, motion analysis, and keyframe matching. Most methods for detecting features rely on pattern analysis of both image and audio processing. Other examples of low-level features include segment breaks, motion flow, on-screen captions and human presence. Low-level features for audio processing include the detection of loud sounds, music, human speech and noise. Language and annotated features provide higher-level information of video content.

Video Interpretation

The next step in multimodal characterization is to interpret video content as a higher-level event. This interpretation is based on how well the low-level video features may correspond to any of a specified set of events or actions. Integrating multiple modalities of video content can result in a more accurate interpretation than by single modality alone. An example is the temporal segmentation of news video described by Hauptmann in 1998 [Hauptmann 1998]. Although shot cuts indicate changes in imagery, audio processing identifies a segment that may include more than a single visual shot.

Figure 1.2 has separated the input to video interpretation modules as "Features" and "Production Standards". They refer to the methods of deriving algorithms. Feature-based experimentation is derived solely from the use of low-level features. The simplest example is image differences that are used to represent shot cuts. Production-based interpretation is derived from standards used in film creation for displaying certain moods or themes. We reverse-engineer our features to fit accordingly. A simple example of this is the use of dissolves and fades rather than shot cuts. Artists use these techniques to indicate a mood change in specific genre of video.

1.2 Browsing Digital Video

Abstraction or summarization of a video is useful in two venues of video presentation: browsing and previewing. Browsing entails viewing a large collection of videos in a short time with minimal loss in content, whereas previewing is defined as viewing a video in its entirety while possibly eliminating portions not essential to content. The application of video browsing is similar to many internet search applications, where a user quickly examines multiple text or video abstracts in order to select an item of interest.

Simplistic browsing techniques, such as increased playback speed and skipping video frames at fixed intervals certainly reduce video viewing time. However, increased video rates eliminate the majority of the audio information and distort much of the image information [Arons 1993], and displaying video sections at fixed intervals merely gives random samples of the overall content. An ideal browser would display only the portion of video pertaining to the content, suppressing less relevant information. Such video, containing only the images pertinent to content, would be considerably smaller than the original and could be used to skim the video in browsing. The audio portion of this video should also consist of only the significant audio or spoken words, rather than simply using the synchronized portion corresponding to the selected video frames.

1.3 Digital Video Libraries

With increased computing power and electronic storage capacity, the potential for large digital video libraries is growing rapidly. These libraries, such as the Informedia™ project at Carnegie Mellon University [Stevens 1994], [Wactlar 1996], [Wactlar 1997], [Christel 1996], and [Christel 2002], will make thousands of hours of video available to a user. Digital video is increasingly used in the World Wide Web and it remains a key component of many educational and entertainment applications. For many users, the video of interest is not always a full-length film. Unlike video-on-demand, video libraries should provide convenient and informational access in the form of brief, content-specific segments as well as full-featured videos. These segments will act as "video paragraphs" for the entire broadcast, allowing the user to view the complete video by moving from one segment to the next.

The information embedded within the digital video must be easy to locate, manage and display. As the size of video collections grows to thousands of hours, potential viewers will need a tool that will help them browse effectively and efficiently. Even with intelligent content-based search algorithms being developed [Mauldin 1991] and [TREC 1993], multiple video segments will be returned for a given query to insure retrieval of pertinent information. The users may often need to view all the segments to obtain their final selections. Instead, the user will want to skim the relevant portions of video for the segments related to their query.

1.4 Characterization and Summarization Research

A multimedia characterization attempts to preserve and communicate the essential content of a video segment via some visual representation. Early work in multimedia abstractions and visualizations include short text titles and single thumbnail images [Zhang 1993] and [Christel 1996].

Another commonly used abstraction is an ordered set of representative, thumbnail images simultaneously on a computer screen. Many researchers addressed the problem of the automatic selection of thumbnails [Arman 1994], [Mills 1992], [Taniguchi 1995], [Rorvig 1993], and [Zhang 1995]. Use of image statistics, such as histogram analysis and texture, camera structure, and shot changes are the dominant methods in these systems. While thumbnails have proven useful in various contexts, their static nature ignores video's temporal dimension. In addition, they often concentrate exclusively on the image content and neglect the audio information carried in a video. Audio actually offers important clues for video.

The video skim was one of the first systems that successfully integrated image, language, and audio understanding for browsing and summarization [Smith 1995] and [Smith 1997]. Other researchers have proposed representations for browsing based on information within the video [Tonomura 1994] and [Zhang 1995]. These systems rely on the motion in a scene, placement of shot breaks, but not on integrated image and language understanding. Video skims are covered in detail in chapters 4 and 5. Other efforts try to combine language and image understanding in order to advantage of multiple modalities. A few notable systems include:

Name-It

This is a system for matching a human face to a person's name in news video [Satoh 1997]. As illustrated in figure 1.5, it approximates the likelihood of a particular face belonging to a name in close proximity within the transcript. Integrated language and image understanding technology make the automation of this system possible.

Spot-it

This is a topological video system that attempts to identify certain characteristics in news video for indexing and classification [Nakamura 1997]. It has reasonable success in identifying common video themes such as interviews, group discussions, and conference room meetings.

Browsing by Clustering

The system [Yeung 1995] was designed to cluster image regions for browsing digital video. It uses many of the image statistics mentioned earlier, but it attempts to process shot transitions rather than just processing individual frames.

News Summarization

Research at AT&T Research Laboratories has shown promising results in video summarization when closed-captions are used with statistical visual attributes [Shahraray 1995]. CNN video is digitized and displayed in an HTML environment with text for audio and a static image for every paragraph.

Audio-Visual Interaction

Another group at AT&T Research Laboratories has performed novel work in audio and image interaction techniques for lip-synchronization in communications [Chen 1995] and [Chen 1997]. Using image-understanding techniques to recognize moving lips and audio understanding techniques to analyze the sound track, their system enhances images and compresses video. This research was continued at the Carnegie Mellon University Advanced Multimedia Processing Lab [Zhang 2002] and [Chen 2002].

Video Abstracts

The Movie Content Analysis (MoCA) group in Mannheim, Germany, has created a system for movie abstraction based on the occurrence of significant image statistics and audio frequency analysis to detect dialogue scenes [Pfeiffer 1996]. This group has also contributed to the field of video characterization with much work in the areas of text detection and segmentation [Lienhart 2000].

High Rate Keyframe Browsing,

The Digital Library Research Group at the University of Maryland conducted a user study to test optimal frame rates for keyframe based browsing [Ding 1997]. They used many of the low-level image analysis techniques, such as color and shot change detection mentioned earlier to extract keyframes. Unlike most research in video browsing, they quantified their work through user studies of a video slide show interface at various frame rates.

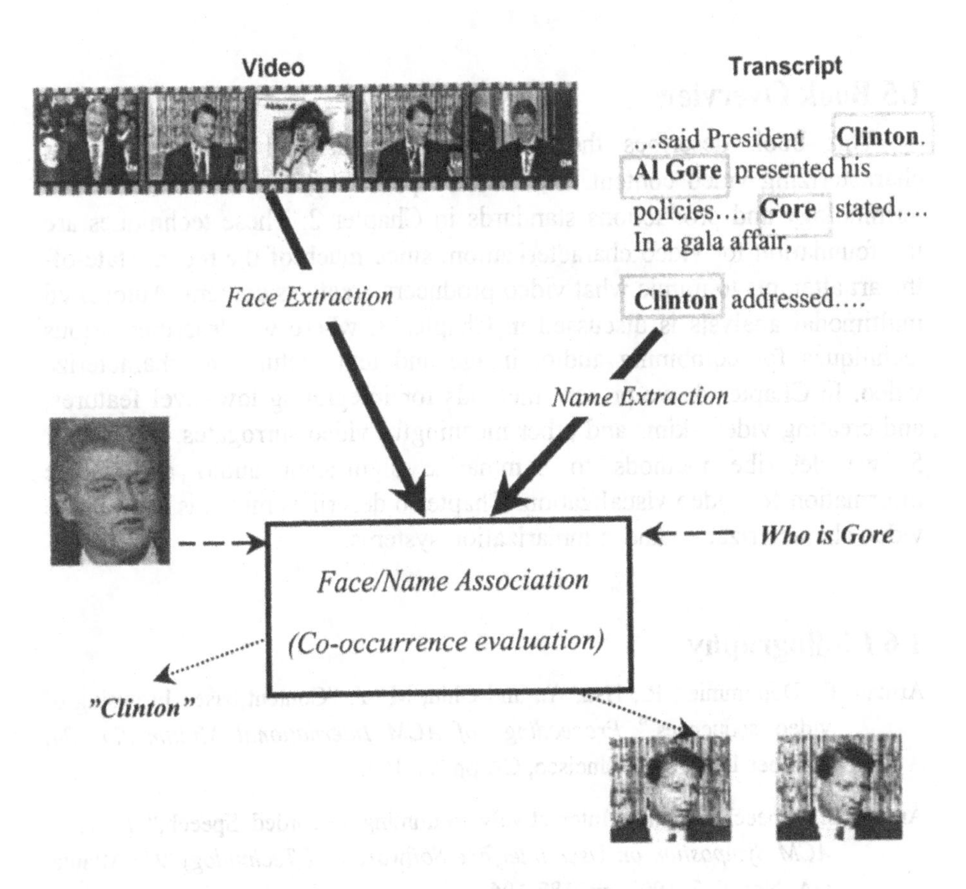

Figure 1.5 Name-it, Face and Name association [Satoh 1997]. Face recognition and identification accuracy is improved through a co-occurrence evaluation with words in the transcript.

Utility Frameworks for Summarization

Sundaram introduced a method to assess visual complexity rather than emphasize multiple forms of semantics [Sundaram 2002]. Much of this work focuses on the complexity of a shot with respect to its duration and proximity to other shots. Recent work includes aspects of audio analysis for summarization.

Video Rhythm

Research in computational media aesthetics [Doria 2002] deals with a history of work in automatic content management and an analysis of video semantics through Film Grammar. A measure for the extraction of a fundamental aspect of film tempo was developed.

1.5 Book Overview

This book describes the current techniques and methodology for characterizing video content. We start by providing an overview of video terminology and productions standards in Chapter 2. These techniques are the foundation for video characterization, since much of the recent state-of-the art attempts to mimic what video producers create as content. Automated multimodal analysis is discussed in Chapter 3, where we describe various techniques for combining audio, image and text features to characterize video. In Chapter 4, we present methods for integrating low-level features, and creating video skims and other meaningful video surrogates. In Chapter 5, we describe methods to summarize significant audio and image information for video visualization. Chapter 6 describes methods to evaluate video characterization and summarization systems.

1.6 Bibliography

Arman, F., Depommier, R., Hsu, A., and Chiu, M. Y. "Content-based browsing of video sequences," *Proceedings of ACM International Multimedia '94*, October 1994, San Francisco, CA pp. 97-103.

Arons, B. "SpeechSkimmer: Interactively Skimming Recorded Speech," *Proc. of ACM Symposium on User Interface Software and Technology'93*, Atlanta, GA, Nov. 3-5, 1993, pp. 187-196.

Bovik, A. "Handbook of Image and Video Processing," Academic Press Publishers, 2000, Boston, MA.

Chen, T., Graf, H. P., and Wang, K., "Lip-synchronization using speech-assisted video processing," *IEEE Signal Processing Letters*, vol. 2, no. 4, pp. 57-59, April 1995.

Chen, T. and Rao, R., "Audio-Visual Interaction in Multimedia Communication" Volume 1, Page 179, *ICASSP*, Munich, 1997.

Christel, M.G., and Pendyala, K. "Informedia Goes To School: Early Findings from the Digital Video Library Project". *D-Lib Magazine*, September 1996.

Christel, M.G., Winkler, D.B., and Taylor, C.R. Improving Access to a Digital Video Library. *Human-Computer Interaction: INTERACT97*, the 6th IFIP Conf. On Human-Computer Interaction (Sydney, Australia, July 14-18, 1997).

Christel, M., Stevens, S., Kanade, T., Mauldin, M., Reddy, R., & Wactlar, H. Techniques for the Creation and Exploration of Digital Video Libraries. Chapter 8 of Multimedia Tools and Applications, B. Furht, ed. Kluwer Academic Publishers, 1996.

Christel, M.G., Hauptmann, A.G., Wactlar, H.D., and Ng, T.D, "Collages as Dynamic Summaries for News Video". *Proceedings of the ACM International Multimedia Conference* (Juan-les-Pins, France, December 1-6, 2002), pp. 561-569.

Cohen, S. "Finding Color and Shape Patterns in Images". Technical Report and Ph.D. Thesis. STAN-CS-TR-99-1620, May 1999.

Ding, W., Marchionini, G., and Tse, T., "Previewing Video Data: Browsing Key Frames at High Rates Using a Video Slide Show Interface." *Proceedings of the International Symposium on Research, Development and Practice in Digital Libraries*, Tsukuba Science City, Japan, November 1997.

Dorai, C. and Venkatesh S., "Media Computing: Computational Media Aesthetics", Editors, Kluwer Academic Publishers, June 2002.

Furht, B., editor, "Multimedia Tools and Applications", Kluwer Academic Publisher, Norwell, MA, 1996.

Furht, B., Editor-in-Chief, "Handbook of Multimedia Computing", CRC Press, Boca Raton, Florida, 1999.

Gonzalez, R. C. and Woods, R. E. "Digital Image Processing". New York: Addison-Wesley, 1992.

Hauptmann, A.G., and Lee, D., "Topic Labeling of Broadcast News Stories in the Informedia Digital Video Library", *Digital Libraries '98 - The Third ACM Conference on Digital Libraries*, Pittsburgh, PA, June, 1998.

Hauptmann, A.G., and Witbrock, M.J., "Story Segmentation and Detection of Commercials in Broadcast News Video", *ADL-98 Advances in Digital Libraries*, Santa Barbara, CA, April, 1998, pp 168 - 179.

Iqbal, Q., and Aggarwal, J. K., "Combining Structure, Color and Texture for Image Retrieval: A Performance Evaluation", *International Conference on Pattern Recognition*, August 2002, pp. 438-443.

Iqbal, Q., and Aggarwal, J. K., "Image Retrieval via Isotropic and Anisotropic Mappings", *Pattern Recognition Journal*. vol. 35, no. 12, pp. 2673-2686, December 2002.

Lienhart, R. and Wolfgang Effelsberg, W., "Automatic Text Segmentation and Text Recognition for Video Indexing". ACM/Springer Multimedia Systems, Vol. 8, pp. 69-81, Jan. 2000

Mauldin, M. "Information Retrieval by Text Skimming," PhD Thesis, Carnegie Mellon University. August 1989. Revised edition published as "Conceptual Information Retrieval: A Case Study in Adaptive Partial Parsing, Kluwer Press, September 1991.

Mills, M., Cohen, J., and Wong, Y.Y. A Magnifier Tool for Video Data. In *Proceedings of the ACM CHI'92 Conference on Human Factors in Computing Systems*. (Monterey, CA, May 1992).

Nakamura, Y., Kanade, T.," Spotting by Association in News Video," *Proceedings of ACM International Multimedia Conference*, November 1997, Seattle, WA.

Pfeiffer, S. Lienhart, R., Fischer, S., Effelsberg, W., "Abstracting Digital Movies Automatically," *Journal of Visual Communication and Image Representation*, Vol. 7, No. 4, pp. 345-353, December 1996.

Rorvig, M.E., "A Method for Automatically Abstracting Visual Documentaries," *Journal for the American Society of Information Science* 44, 1, 1993.

Rubner, Y., C. Tomasi, C., and Guibas, L. J. "A Metric for Distributions with Applications to Image Databases". *Proceedings of the 1998 IEEE International Conference on Computer Vision*, Bombay, India, January 1998, pp. 59-66.

Satoh, S., Sato, T., Smith, M., Nakamura, Y., Kanade, T., "Name-It: Naming and Detecting Faces in News Video" *Computer Vision and Pattern Recognition*. San Juan, PR, June 1997.

Santini, S., Jain, R., "Similarity Matching" *IEEE Transactions on Pattern Analysis and Machine Intelligence*, 21(9): 871-883, 1999.

Shahraray, B., Gibbon, D.C., "Automated Authoring of Hypermedia Documents of Video Programs," *Proceedings of the Third ACM International Multimedia Conference*, pp. 401-409, San Francisco, CA, November, 1995.

Shahraray, B., and Gibbon D.C., "Automatic Generation of Pictorial Transcripts of Video Programs" *Multimedia Computing and Networking 1995, Proc. SPIE* 2417, February 1995.

Shapiro, L.G. and A. Rosenfeld, "Computer Vision and Image Processing", Academic Press Publishers, 1992, Boston, MA, edited volume.

Smith, M.A., and Kanade, T., "Video Skimming for Quick Browsing Based on Audio and Image Characterization." Carnegie Mellon technical report CMU-CS-95-186, July 1995.

Smith, M., Kanade, T. "Video Skimming and Characterization through the Combination of Image and Language Understanding Techniques". *Computer Vision and Pattern Recognition (CVPR)*. San Juan, PR, June 1997.

Stevens, S., Christel, M., and Wactlar, H. "Informedia: Improving Access to Digital Video". *Interactions* 1 October 1994, pp. 67-71.

Sundaram, H., Xie, L., Chang, S. "A Utility Framework for the Automatic Generation of Audio-Visual Skims", *Proceedings of the ACM International Multimedia Conference*, Juan Les Pins, France, Dec.1-6, 2002.

Taniguchi, Y., Akutsu, A., Tonomura, Y., and Hamada, H. An Intuitive and Efficient Access Interface to Real-Time Incoming Video Based on Automatic Indexing. *Proceedings of the ACM International Multimedia Conference*. (San Francisco, CA, November 1995).

"TREC 93," *Proceedings of the Second Text Retrieval Conference*, D. Harmon, editor, sponsored by ARPA/SISTO, August 1993.

Wactlar, H.D., Kanade, T., Smith, M.A., and Stevens, S.M. "Intelligent Access to Digital Video: Informedia Project," *IEEE Computer*, 29, May 1996, 46-52.

Wactlar, H., Hauptmann, A., Smith, M.A., Pendyala, K., Garlington, D. "Automated Video Indexing of Very Large Video Databases," *SMPTE Journal*, August, 1997.

Yeung, M., Yeo, B., Wolf, W., and Liu, B., "Video Browsing Using Clustering and Scene Transitions on Compressed Sequences". *Proceedings IS&T/SPIE Multimedia Computing and Networking*, February 1995.

Zhang, C., and T. Chen, T., "An Active Learning Framework for Content Based Information Retrieval", *IEEE Transaction on Multimedia*, Special Issue on Multimedia Database, June 2002.

Zhang, H., Kankanhalli, A., and Smoliar, S. "Automatic partitioning of full-motion video," *Multimedia Systems* 1993 1, pp. 10-28.

Zhang, H.J., Smoliar, S., and Wu, J.H., "Content-Based Video Browsing Tools," *Multimedia Computing and Networks,* Multimedia Systems, 2, 6 (1995), 256-266.

Zhang, H.J., Tan, S., Smoliar, S., and Yihong, G. "Video Parsing, Retrieval and Browsing: An Integrated and Content-Based Solution," *Proceedings of the ACM International Conference on Multimedia,* San Francisco, CA, November, 1995.

Chapter 2

Video Structure and Terminology

Digital video contains a complex assortment of audio, visual, and often, textual information. Production standards have been developed to effectively convey a variety of messages and themes in video. Producers and directors often employ these "tools of the trade" in video creation and editing. Their roots first appeared in analog media, and they represent a long history of knowledge in how production video is structured. The potential use of that knowledge for automated video characterization through image, audio and language processing systems is discussed in detail throughout this chapter. This chapter provides an understanding of video from a producer's perspective, which is necessary to develop the automated characterization systems described in subsequent chapters.

Section 2.1 of this chapter defines video terminology. Section 2.2 describes various categories of video content. Section 2.3 describes video production and editing standards, and their relationship to video characterization. Section 2.4 is an introduction to video summarization as a manual process. This section also describes existing characterization, summarization and segmentation systems in use today.

2.1 Video Terminology

This book uses common terminology from production standards, and image, audio and video processing technology. To help understand the notation, we first provide introductory definitions of some video terms used in this book. Figure 2.1 illustrates the video shot, scene, frames and audio track.

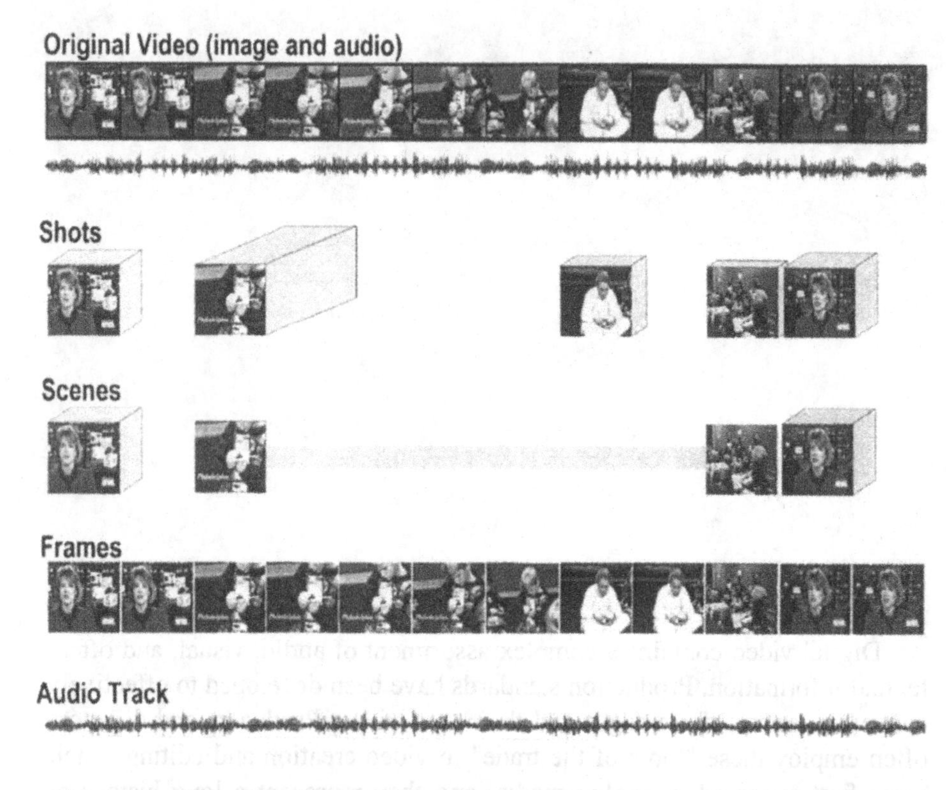

Figure 2.1 Illustration of video terminology: Original Video, Shots, Scenes, Image Frames and the Audio Track. The "Original Video" contains images sampled from a CNN News story on the K'nex Toy (December 1994). Shots - represent cuts, fades, wipes or other forms of abrupt changes in imagery. Scenes contain a consistent thematic unit. The Z-axis shown for shots and scenes represents the time axis. The three scenes above depict the following elements: 1.) The introduction of the news story, 2.) The actual story with on-sight interviews, and 3.) The closing remarks and transition to another story. The frames and audio represent the image and audio track, respectively.

Term (modality)	Definition
Video-Feed-Stream (audio and image)	The terms video, feed, and stream will be used to represent a sequence of images and audio. These terms will seldom be used to describe the image track without audio. During production or broadcast, the "feed" is often described separately as an "audio feed" or "video feed".
Full-Length Video	The term "full-length" is used to describe a video that is produced professionally with studio quality. This usually implies feature films, documentaries, or news programs.
Shot (image only)	This term defines a single camera shot. It is also used to describe change or borders of image content. Cuts, Fades, Dissolves and Wipes, for example, delineate shots. [Zhang95]
Frames (image)	Frames refer to the actual image portion from the image track of the video.
Show-Story-Segment (image and audio)	The term show, story, or segment may be used interchangeably, but in all cases, they represent a video that is independent of other content. Shows are typically shown on broadcast television for home use. An example might be a common sitcom, movie, talk show, etc. A segment or story usually represents a smaller portion of a show, such as a single topic in a full-length news broadcast.
Scene (image and audio)	A scene is a subset of the video that makes up a semantic unit. This unit may consist of several shots and phrases.

Audio Track (audio) — The audio track refers to the actual audio stream in the video. This may be compressed, although many applications still use uncompressed WAV files for faster analysis.

Word (audio or text) — A word is a portion of the audio signal or text transcript which represents a single word.

Transcript (text) — The transcript is an ANSII representation of the spoken words in an audio or video signal. It is usually provided through speech recognition, closed-captions, production notes, or an actual script. The transcript may be included in production notes, and it generally refers to that part of the audio that is spoken. Some production notes contain information about sound and imagery effects.

Phrase (audio or text) — A phrase is a collection of words that make a semantic unit of audio or text. A phrase may bound the range of a sentence, but this is not a necessary condition. A sub-region of a sentence may also serve as a phrase.

The notation for video timing and duration follows a common format; hours, minutes, seconds, and frames are listed respectively. An example is:

Hours (0-N): Minutes (0-59): Seconds (0-59): Frames (0-29)

Examples	**0:08:56:15**	8 minutes, 56 seconds, 15 frames
	1:22:10:00	1 hour, 22 minutes, and 10 seconds

Video Storage Mediums

Digital video compression schemes offer increased storage capacity and statistical image characteristics, such as DCT (Discrete Cosine Transform) filtering coefficients and motion compensation data. For example, the DCT coefficients preserve colors, texture, and other spatial domain characteristics, and motion vectors preserve object motion, camera pan and zoom, and other temporal characteristics.

One drawback to video compression is loss in quality. Bitstreams created by lossy compression schemes preserve statistical information, but they produce visible artifacts with high compression. Lossless schemes, such as Run Length Encoding (RLE) and Huffman coding, do not sacrifice quality but provide lower compression ratios. Furthermore, bitstreams created by lossless algorithms do not explicitly contain any statistical information of the original video.

Many compression algorithms provide compression as high as 100 to 1, depending on resolution, and often use DCT and motion compensation for compression. The DCT parameters may be used for video segmentation while the motion compensation statistics may be used as a form of optical flow, as discussed in Chapter 3.

Storage capacity and data access are important in image and video retrieval systems. Data can be stored in a hard drive, CD-ROM, DVD, Digital Tape or many other storage mediums. The time needed to access information on a particular medium will fluctuate and so retrieval algorithms must be flexible. The storage requirements will also change depending on the size and quality of video. An hour of video at full SIF (352x240 in NTSC) constitutes approximately 640 Megabytes with the MPEG-1 compression standard.

2.2 Video Categories Used in this Book

It is important to understand that most video can be categorized into specific groups. We may exploit these categories later for more precise characterization. The experiments in this book describe methodology and experimentation with documentaries and broadcast news, with some experimentation devoted to sports and feature-films. The video data used in this book is discussed below:

2.2.1 Documentaries

Documentary footage follows many standard video production procedures. Access to public broadcast material is less stringent than private material since much of what they produce is educational, rather than commercial. There are three basic formats for documentary video: Factual, Historical, and Biographical.

Most factual documentaries follow a single theme and build on that theme with supportive segments. Many of the factual documentaries are based on scientific topics. Historical documentaries usual focus on a single event. They often describe the event through flashback and present day commentary. Biographical documentaries follow a similar structure as the historical documentary, but will often include recorded footage of the subject. Our experiments studied several hours of documentary video in conjunction with research for the Informedia Digital Video Library Project [Wactlar 1997] and [Wactlar 1996]. Below is the list of the video contributors to the figures in this book and the associative research:

- WQED Pittsburgh - A Public Broadcasting Station (PBS) in Pittsburgh, PA, with a collection of local programming and documentaries. Notable topics include: Space Exploration, Geology, Zoology, Materials Science, and Oceanography.
- Open University - This is a collection of educational videos in Math, Physics, Biology, and Engineering, provided by the Open University, United Kingdom.
- U.S. Geological Survey - A large collection of public domain video is available through this agency. Topics include: Plate Tectonics, Transportation, Ecology, and Anthropology.
- Smithsonian Institute - This is a large collection of scientific and historical documentaries from the Smithsonian Museums.

2.2.2 Broadcast News

The corpus for experimentation with news video includes pre-recorded and live footage from CNN Headline News, ABC, NBC, and CBS. Most live news segments present a variety of challenging editing styles and effects that are not commonly used in pre-recorded segments. Differences in methodology are needed for segmentation and summarization depending on the news corpus. News video is best categorized into the following groups:

- Live News - News that appears live and unedited. Anchorpersons narrate segments of news footage from a written script. Edited news stories may be inserted at specific time intervals. The script and relative consistency of the production format may be used to better characterize the video. In later sections, we describe video-logging systems that automatically annotate a video feed. In most cases, the desired output is a near real-time annotation of the input video. These systems often require dedicated hardware or optimized software for fast processing.
- Pre-recorded News – News that has been recorded and edited. This format is seldom used, as news is a changing media that requires constant updates. In some cases, a news station will interchange pre-

recorded video segments with live video footage. This is the model used by such networks as CNN, where new video segments are inserted with older stories. Pre-recorded news may not have the same processing constraints as Live News. When the same segment is shown multiple times, the requirement for real-time processing is diminished. The process of multiple editing also provides additional manual annotation to the video.

2.2.3 Feature-Films

The films mentioned in this book were commercially successful films suitable for all audiences. Most of them were produced after 1950 to ensure video quality and subject knowledge for user-studies and experimentation.

- Action – Films that contain action sequences are popular mediums for video analysis. In many cases it is quite easy to identify an area of interest based on specific audio parameters or an increase in camera or object movement. When an area of interests does not include one of these elements, characterization is both difficult and unpredictable. One producer may use silence to indicate heightened action, where another will use a loud explosion.

- Comedy – Films with comedic themes are extremely difficult to analyze. This is caused, in part, to the difficult nature in creating a laughing response. Not all "funny" scenes induce the same type of laughter. Some forms of human laughter can be modeled automatically [Stolcke 1996].

- Romance – This is also a very difficult theme to analyze. Like the comedic theme, it is difficult to determine which scenes in a film induce romance.

- Drama and Suspense – Although dramatic films often contain plot twists and complex themes, they typically follow standard formats and procedures. Many producers use the same "tricks of the trade" to induce suspense or climax. As these procedures follow a conventional format, it is often possible to characterize certain scenes. For example, music or loud sounds without dialog are often used to convey tension or a transitional period. This is true for many popular films, such as The Godfather, Star Wars, E.T., Jurassic Park, and a host of other films.

- Adult Film – It is worth mentioning that many researchers are developing systems to detect the presence of human skin [Jones 1999]. One goal of their work is to identify imagery that might contain inappropriate content for children on the internet and in video.

Figure 2.2 Different playing surfaces for outdoor sports: (Upper Left) Close-up of grass image, (Upper Right) Grass field with subtle dirt patches, (Bottom Left) Close-up of artificial turf, and (Bottom Right) Artificial turf field with consistent color and texture.

2.2.4 Sports

Most sporting events can be grouped into two types: 1.) Play-based sports and 2.) Continuous action sports. Video characterization of play-based sports is easier than continuous, as the duration of each "play" is short and well structured in sports such as football and baseball. In continuous action sports, like soccer and basketball, each play tends to last longer with less clear start or end demarcations.

Play-based Sports

- Football – Football is filmed in many formats that make it difficult to analyze. Most plays start with the quarterback initiating a play, but the angle at which the shot takes place can change dramatically. Most football events take place outdoors, where the time-of-day and changes in weather can affect lighting. Football also has a wide array of plays, including the pass, run, option and turnover. Unlike the other play-based sporting events, a change of possession is possible during a turnover, which essentially reverses the directional flow of the game. Many applications, such as advertisement insertion and subject tracking, require that the player be separated or extracted from the field of view. Artificial turf provides a uniform static background that is better for computer vision and image processing techniques. Figure 2.2 shows examples of grass and turf textures. The grass is more complex with varying levels of texture and color. Artificial turf maintains a more consistent background.

- Baseball – Video from baseball is straightforward to analyze and researchers have had success in identifying plays [Rui 2000], [Chang 2002], and [Pan 2002]. This is due to its fixed structure. A hitter at home plate appears in the same pose as all other hitters. Unlike football, there is no turnover, and plays usually fall into a limited number of categories, such as, a single, home run or foul.
- Sumo Wrestling – This play-based sport is also straightforward to analyze [Escamilla 2000] and [Li 2001]. The display is rigid and most action is based on human motion. Figure 2.3 shows a test image for a system that was designed to detect the start and end of a Sumo match based on occurrences of object motion. The system could not distinguish human motion, but the display format in a Sumo match is so rigid that all object motion comes from the wrestlers.

Figure 2.3 Sumo Wresting Match. Wrestlers contribute to most movement and usually remain in the field of view. In this sport, the detection of the start of any action is sufficient to identify the beginning of the match. Most matches end in less than 20 seconds, so there is no need to segment the video beyond identifying the start of some action or moving objects.

- Tennis – In tennis, a camera is usually positioned behind the player receiving the serve. This provides the best viewing angle as well as a rigid framework for video analysis. Plays may be detected by tracking the start of play through human movement, ball movement or audio clues. Researchers have also tracked ball movement for interactive television applications [Pingali 2000] and [Sudhir 1998]. Figure 2.4 shows an interactive ball tracking system developed by Lucent Technologies.

- Volleyball – This sport is difficult to analyze due to the continuous movement of the camera. The camera angle is usually centered at the net to allow for a maximum court view. This angle also requires that the camera follow the ball back and forth across the net to view action close-up. This constant movement, coupled with inconsistent indoor or outdoor lighting, provides a poor basis for video analysis.

- Track and Field – These events take place over an entire day, so the actual segments are well edited prior to viewing. Many researchers focus on human motion understanding rather than the video characterization aspects [Ju 1998], [Aggarwal 1999], and [Metoyer 2000]. The utility of these systems becomes diminished with the recent advances in moving cameras. Cameras are now positioned along the tracks to follow the athletes during an event and as they cross the finish line.

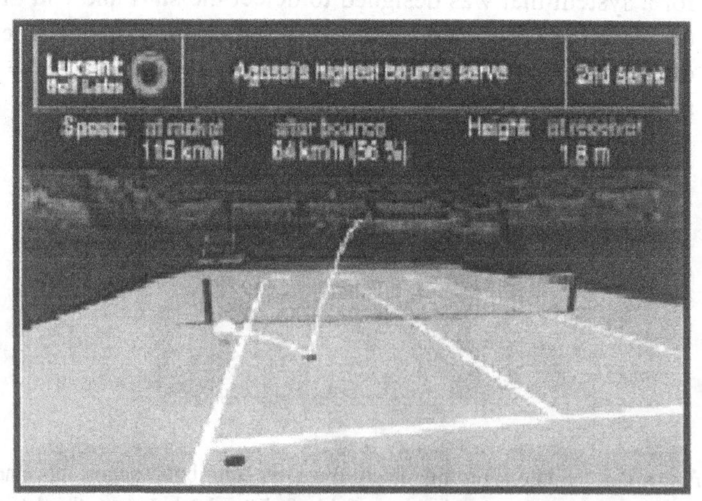

Figure 2.4 Lucent Technologies – Enhanced Viewing System for interactive tennis. The interface provides playback of individual scores and tracks movement and ball placement throughout the match. A tennis ball can be tracked on most court surfaces [Sudhir 1998].

Continuous Action Sports

- Soccer – This sporting event is difficult to characterize because of the continuous movement and the relatively small number of scores per game. Players move continuously between changing camera shots. The excitement associated with an impending score is similar in image and audio characteristics of an actual goal. In both cases, the audience noise is quite high followed by a pause in activity. Even with the difficulties in analyzing this type of content, researchers have achieved some success towards summarization of soccer content [Ekin 2003] and [Yow 1995].

- Basketball – Basketball has the benefit of periodic pauses in play during scores and "fouls". Scoring occurs quite frequently, and the type of score determines the highlights in a game rather than the margin of victory. Zhou describes a method for indexing basketball games based on a set of rules [Zhou 2000]. The "dunk" and 3-point shot are more exciting than a conventional shot. In many cases these highlight shots are quickly followed by another play and cannot be detected by audio reaction. Potential methods for detecting dunks could include proximity detection based on the position and velocity of a human hand with respect to the rim. The 3-point play can be detected by manually or automatically monitoring changes in the score. This might require a system for detecting and recognizing on-screen text if the scores are not available on-line.

- Ice Hockey – The relatively constant surface in ice hockey provides an ideal background for such operations as player segmentation and tracking. The constant surface is ideal for puck tracking systems that are used in many broadcasts to highlight the puck against the ice. Although a team uniform may be white or bright colored, it must also contain sufficient dark coloring to enable player recognition during viewing. Like soccer, scores do not occur with great regularity. An impending or failed scoring attempt will have similar image and audio characteristics to an actual goal.

Sports Replay

Although play-based sports and continuous action sports differ in format, they both present highlights shortly after a play of interest. In most cases, this highlight is shown in slow motion. Slow moving video usually produces higher quality optical flow, but the resolution and frame rate for slow motion video is often the same as conventional video. This results in a blurred or low-resolution image, which is not optimal for optical flow. Figure 2.5a shows the blur or smearing effect that takes place when motion video is frozen or played at a slow rate.

An analysis of the optical flow can be used to detect slow motion playback and identify the previous scene as a highlight [Kobla 1999] and [Rees 1998]. There are many working systems detecting optical flow, and section 3.3 provides an overview of this technology.

Figure 2.5b shows the resulting motion vectors from Kobla's work on slow motion replay. In this work, the motion vectors used for compression are analyzed for slow motion detection.

Figure 2.5a Motion blur and low-resolution imagery from an American football game. The image is distorted and difficult to analyze with conventional optical flow [Kobla 1999].

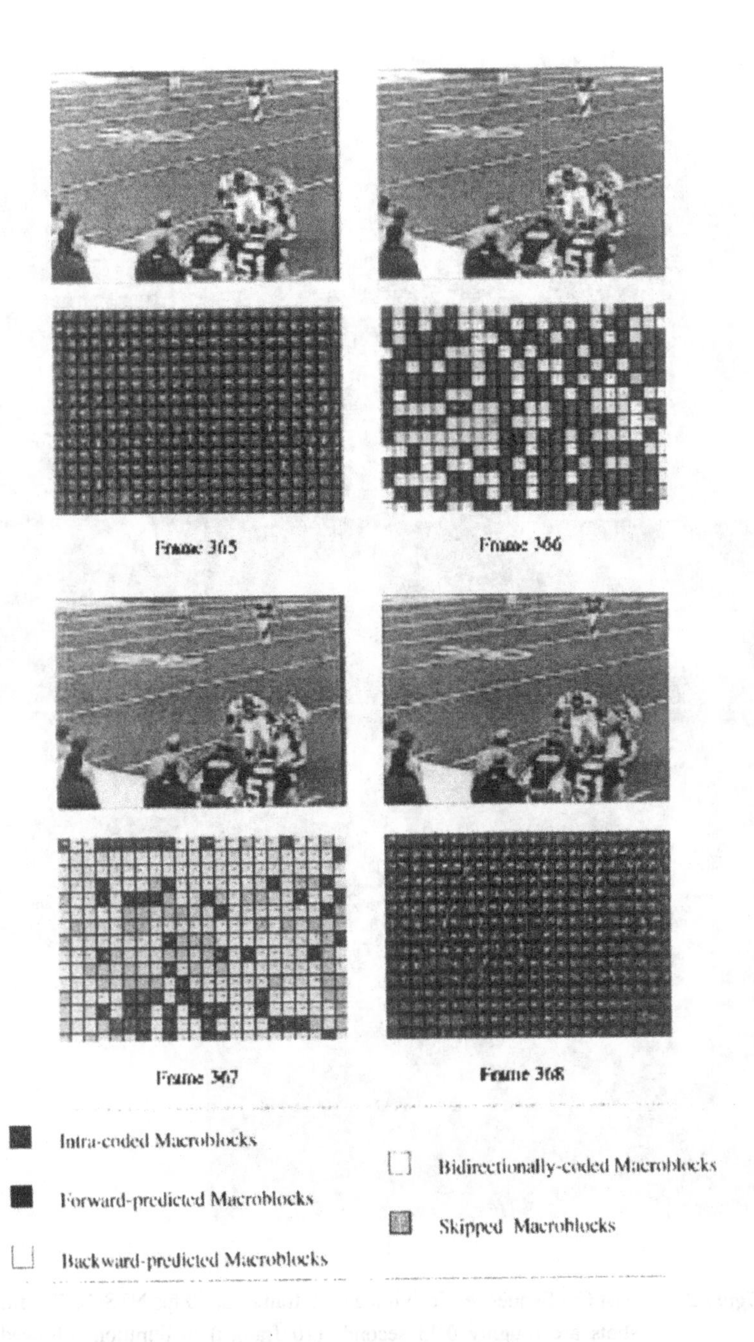

Figure 2.5b Image sequence of slow motion replay and corresponding MPEG frames [Kobla 1999].

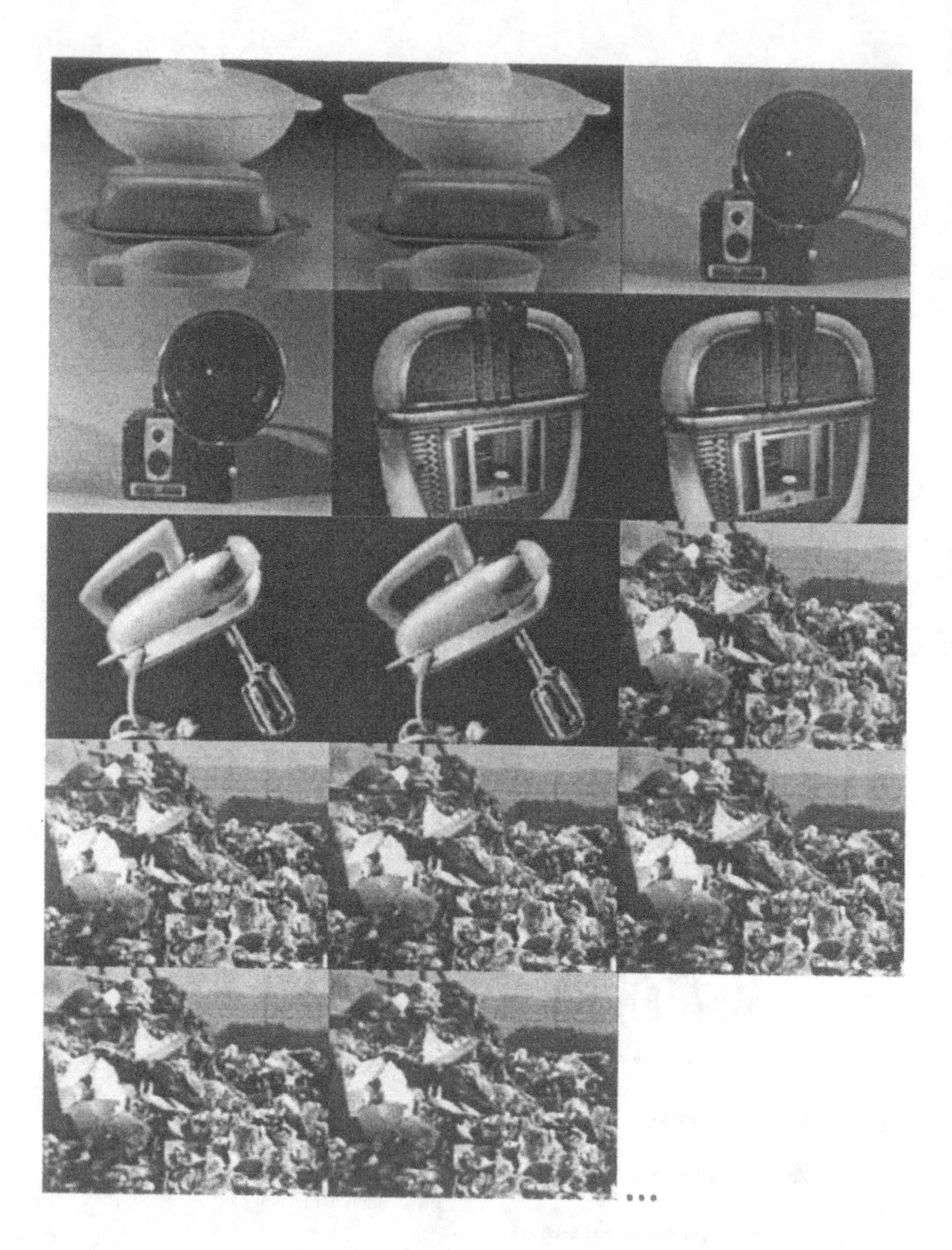

Figure 2.6 Fast Cut Sequence (Each image = 5 frames at 30 fps NTSC). The first four shots are roughly 0.33 seconds (10 frames) in duration followed by a much longer shot.

2.3 Video Production and Editing Standards

Video production manuals provide insight into the procedures used during video editing and creation [Bordwell 1993], [Millerson 1990], [Millerson 2001], [Pryluck 1982], and [Smallman 1970]. One of the most common elements in video production is the ability to convey climax or suspense to an audience. Producers use a variety of different effects, ranging from camera positioning, lighting, and special effects for that purpose. Detection of such elements in video is beyond the realm of present image and language understanding technology. However, a small subset of basic editing techniques can be automatically detected for video understanding.

2.3.1 Video Cuts - Shot Changes

There is a distinction in terminology between a shot and a scene. A scene represents a semantic unit and may be a collection of one or more shots. The boundaries for a scene change are subjective and may differ from one user to the next. Shot boundaries are more definitive, as they separate specific visual content in video.

The most fundamental shot change is the video cut. For a cut, the static difference between image frames is so distinct that accurate detection is not difficult. There are a variety of more complex shot changes used in video production, but the basic premise is a change in visual content. The video cut, as well as other shot change procedures is discussed below.

1. Fast Cut - A sequence of video cuts, each very short in duration, represents a fast cut. This technique heightens the sense of action or excitement. This procedure is used intermediately in documentaries and often in feature-films. To detect a Fast Cut, we may look for a sequence of shot changes that are in close proximity. An example of this is shown in Figure 2.6 where many different samples of plastic appliances are shown at a rapid pace to introduce a darker element of waste disposal (Infinite Voyage, WQED, Pittsburgh, PA).

2. Distance Cut - A distance cut occurs when the camera cuts from one perspective of a shot to another some distance away. This shift in distance usually appears as a cut from a wide shot to a close-up shot, or vice-versa. They help to clarify the content and location of a shot over time and allow the producer to control the exact perception of the audience. An example of a distance cut is shown in Figure 2.7, where a scientist is first seen in a close-up followed by a distance cut to a lab environment (Infinite Voyage, WQED, Pittsburgh, PA). This procedure

is often used in documentaries to illustrate a particular concept or idea. It is used in feature-films for a broader variety of themes and settings.

The detection of this procedure is not straightforward. It requires precise object registration and recognition at varying size, lighting, orientation and position.

Figure 2.7 Distance Cut with an initial close-up shot of a researcher followed by a zoom-out shot of the researcher and his oscilloscope. The distant cut is difficult in most settings. The subject will often undergo some form of rotation, translation, or object occlusion. A distance cut may be detected when the scale is generally on the order of 0.7X or less.

3. Intercutting - When shots change back and forth from one subject to another, we say the subjects are intercut. This concept is similar to the distance cut, but the images are separate and not inclusive of the same shots. Intercutting is used to show a thought process between two or more subjects. It is most often used as an element of tension or suspense

in feature-films. Figure 2.8 shows an example of intercutting between two human subjects (Infinite Voyage, WQED, Pittsburgh, PA). Intercutting appears in documentaries and feature-films, and is quite popular in news footage during interviews or commentaries.

Intercutting is possible to detect when there is little change between the subject shots, which is usually the case in most facial intercuts. The histogram matching techniques described in chapter 3 may be used to adequately detect this procedure.

4. Dissolves and Fades - Dynamic imaging effects are often used to change from one shot to another. A common effect in all types of video is the fade. A fade occurs when a shot changes over time from its original color scheme to a black background. This procedure is commonly used as a transition from one topic to another. Another dynamic effect is the dissolve. Similar to the fade, this effect occurs when a shot changes over time and morphs into a separate shot. This transition is less intrusive and is used when subtle change is needed. An example of fade is shown in Figure 2.9a, where a single gem set against snow changes to a larger set of gems (Planet Earth II, WQED, Pittsburgh, PA). An example of the dissolve is shown in Figure 2.9b, with a black background that changes to an image of two helicopters flying over mountains (Infinite Voyage, Mass Extinction, WQED, Pittsburgh, PA). This sequence is later used in section 3.4 (Figure 3.6) to illustrate histogram changes in a dissolve sequence.

   We can detect these effects by examining the behavior of their color histogram overtime [Hampapur 1995]. Each effect is approximately linear, so automatic detection is plausible. There are techniques for detecting most categories of shot changes. Several researchers have published surveys on this topic in recent years [Boreczky 1996], [Gargi 2000], and [Lienhart 2001].

Figure 2.8 Intercutting sequence between a human subject's face and a close-up of his hands holding a knife and a pair of scissors.

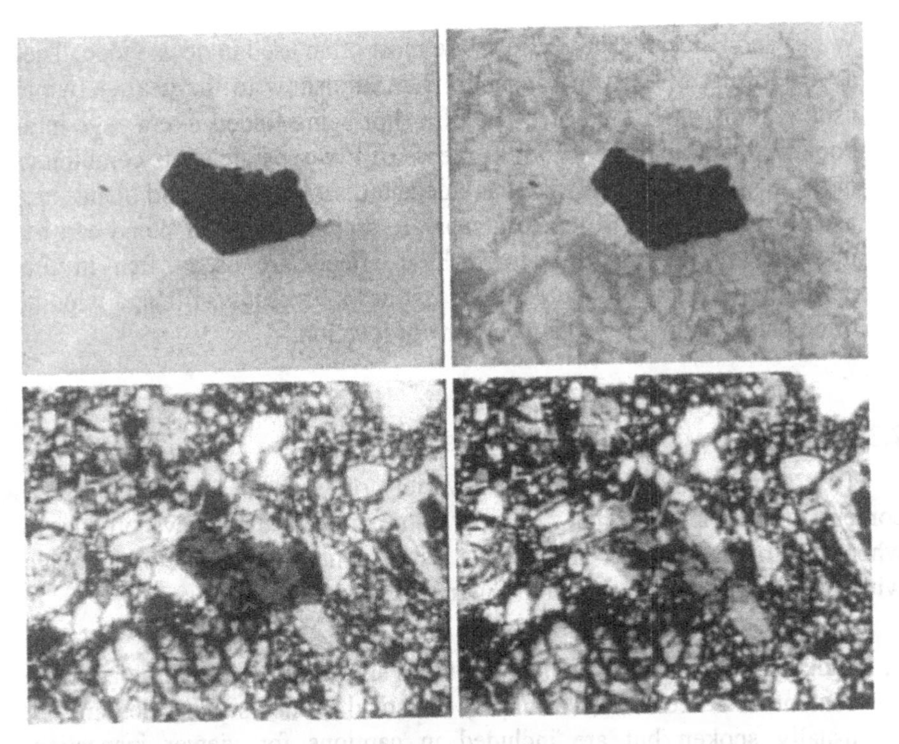

Figure 2.9a Fade sequence.

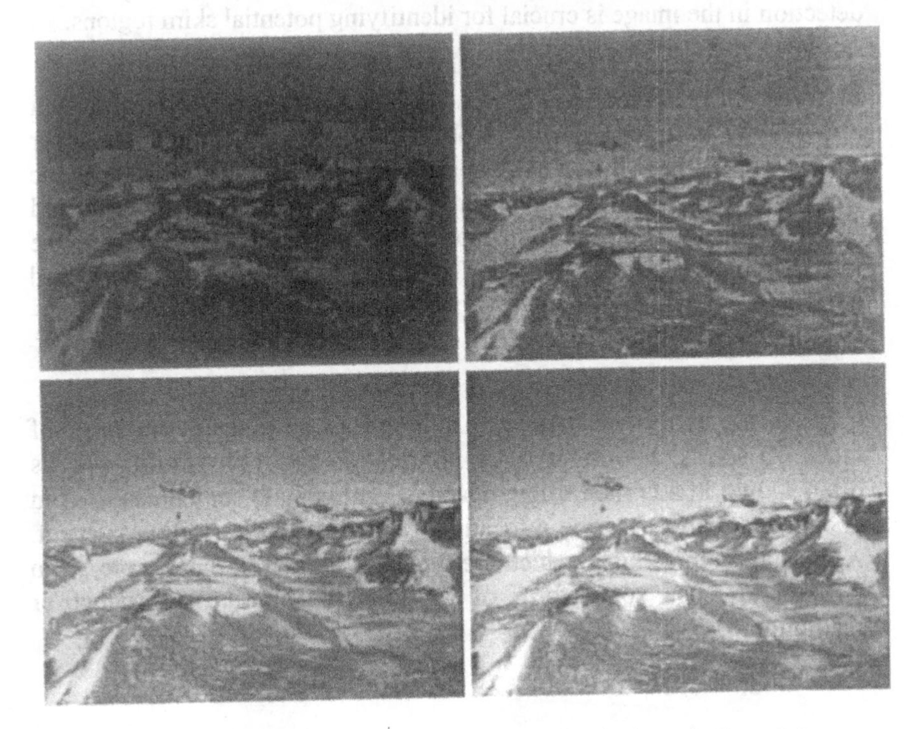

Figure 2.9b Dissolve sequence. Histogram analysis shown in figure 3.6.

5. Wipes and Blends - These effects are most often used in news video. The actual format of each may change from one show to the next. A wipe usually consists of the last frame of a shot being folded like a page in a book. A blend may be shown as pieces of two separate shots combining in some artistic manner. Like the fade and dissolve, wipes and blends are usually used for transition to a separate topic. A wipe or blend can be inserted in real-time; therefore, these effects are used often in the production of live and edited broadcast news. In feature-films, a wipe is often used to convey a change in time or location.

2.3.2 Video Captions and Graphics

Text and graphics are used in a variety of ways to convey the video content to the viewer. They are most commonly used in news broadcast, where information must be absorbed in a short time. An example of an image with text and graphics is shown in Figure 2.10.

1. Video Captions - Text in video provides significant information as to the content of a scene. For example, statistical numbers and titles are not usually spoken but are included in captions for viewer inspection. Moreover, this information does not always appear in closed captions so detection in the image is crucial for identifying potential skim regions.

   In news video, captions of the broadcasting company are often shown at low opacity in a corner without obstructing the actual video. A ticker tape is widely used in news broadcast to display information such as the weather, sports scores, or the stock market. In Figure 2.10, the time and region is displayed in a ticker-tape format with the news logo in the lower right corner at full opacity. Captions that appear in the lower third portion of a frame are usually used to describe a location, person of interest, title, or event in news video.

   Captions are used more frequently in broadcast news than other types of video. In sports, a score or some information about an ensuing play is often shown in a corner at low opacity. Captions are sometimes used in documentaries to describe a location, person of interest, title, or event. Almost all commercials, which are typically limited in duration to 30 to 60 seconds, use some form of captions to describe a product or institution.

Figure 2.10 News anchor with supportive graphics that delineate a subject change. The position of the graphic may change depending on the news broadcast.

For feature-films, a producer may use text at the beginning or end of a film for deliberate viewer comprehension, such as character listings or credits. A producer may also start a film with an introduction to the story. Throughout a film, captions may be used to convey a change in time or location, which would otherwise be difficult and time consuming to display. A producer will seldom use fortuitous text in the actual video unless the wording is noticeable and easy to read in a short time. One exception is the intentional use of product placement for advertisement.

A typical text region is a horizontal rectangular structure of clustered sharp edges, because characters usually form regions of high contrast against the background. By detecting these properties, we can extract potential skim regions from video frames that contain textual information. Most captions are high contrast text such as the black and white chyron commonly found in news video. Consistent detection of the same text region over a period of time is probable since text regions remain at an exact position for many video frames. If the text of interest is fixed in a specific location, a time-based average may correct false detections that occur when moving captions or special effects are placed in a shot. Procedures for text detection are discussed in chapter 3.

For some fonts a generic optical character recognition (OCR) package may accurately recognize video captions. For most OCR systems, the input is an individual character. This presents a problem in digital video since most characters degrade during recording, digitization and compression. Sato *et al* [Sato 1998] presents a technique to improve the quality based on time average and up-sampling. For a simple font, we can search for blank spaces between characters and assume a fixed width for each letter.

2. Graphics - A graphic is usually a recognizable symbol that may contain text. Graphic illustrations or symbolic logos represent a variety of institutions, locations, and organizations. A logo representing the subject is often placed in a corner next to an anchorperson during dialogue. Detection of graphics is a useful method for finding changes in semantic content; its appearance may serve as a scene break. Recognition of corner regions for graphics detection may be possible through an extension of the shot change detection technique, such as histogram analysis of isolated image regions instead of the entire image. Procedures for graphics detection are discussed in chapter 3.

2.3.3 Motion Video

Procedures in camera motion are used in all types of video to convey a variety of visual effects [Millerson 1994]. Camera positioning, movement and focus may be altered to isolate regions of importance or to simply show transition.

1. Pans - A pan occurs when the camera moves in a single direction for an extended period of time. This effect is used in varying degrees with all types of video. In documentaries, it is often used during a transition from one segment to the next. This is usually the case when the camera is panning outdoor footage covering an area of continual imagery. Examples of this type of pan can range from a single horizontal shot of open scenery to a ground-to-head vertical pan of a human speaker. An indoor pan is typically used to view a variety of different objects and is seldom used for transition.

Using the optical flow analysis described in chapter 3, we may approximate areas of panning camera motion, which generates constant and unidirectional flow. An example of a pan is shown in Figure 2.11(A), where the camera moves to the right during a transition scene over a landscape (Infinite Voyage, Human Archeology, WQED).

2. Head-on Shot - This occurs when the camera is partially fixed, but a subject moves continually towards the center of view. When the motion is slow, these shots are used primarily for transitions. Transitional head-on shots are used in documentaries, feature-films, and broadcast news. Fast motion implies suspense and a potentially important scene for summarization. Fast head-on shots are commonly used in films and during replays of sports video.

   We can locate this type of shot with camera and object motion detection. The optical flow for a head-on shot exhibits characteristics similar to object motion or a zoom.

3. Single Object Motion - This effect is quite similar to the head-on shot, with the exception that the object motion is not limited to the direction of the camera. Single object motion is used often and in all types of video. Detection of this effect is possible through the object detection procedures described in chapter 3. It is easier to detect whether an object is moving or stationary than to recognize or track an object. Figure 2.11(B) shows an example of a single subject walking (Infinite Voyage, Advanced Materials, WQED).

4. Stationary Subject or Rotating Camera - This procedure is defined by a camera moving around a stationary subject. Usually the subject is the focal point of a scene. In documentaries, it is often used to show many aspects of an intricate or large object. It may serve the same purpose in feature-films or as a special effect. The same effect is used by Motion Analysis Inc. and ESC Entertainment Inc. for the Matrix movies and by EyeVision for football Superbowl broadcasts [Collins 2002].

   Figure 2.11(C) is an example of this technique used in a documentary to show the Golden Gate Bridge, San Francisco, CA, from an aerial view (Planet Earth II, WQED). This procedure may be identified through camera motion analysis. Its detection is a potentially important element in video summary creation because it is seldom used and it usually displays an object of extreme importance.

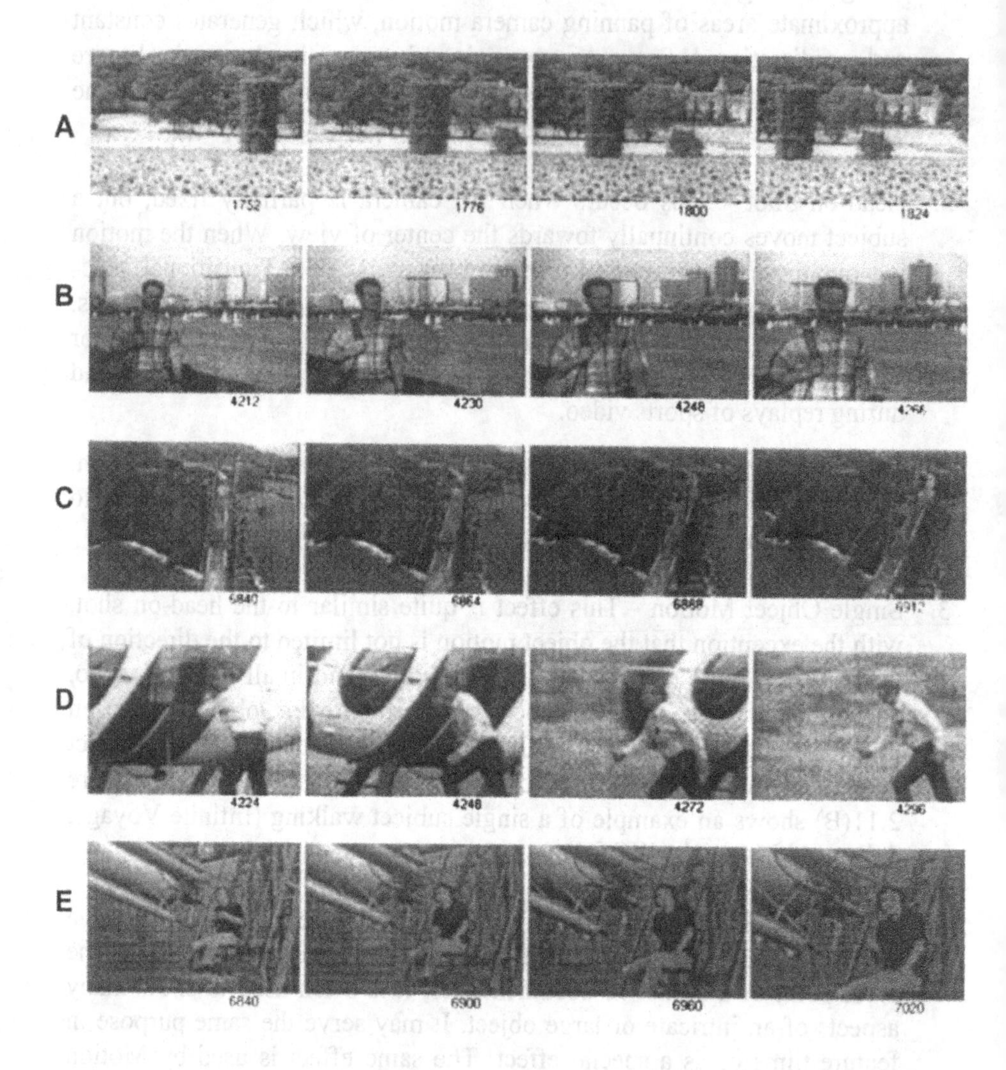

Figure 2.11 Examples of various motion video shots: (A) Panning sequence from right to left, (B) Single object motion – camera remains fixed while a moving subject walks from left to right, (C) Stationary subject with rotating camera, shot from an aerial view, (D) Tracking shot, with a helicopter landing (not shown) and a human subject walking (E) Zoom-in shot of a subject sitting on a telescope. The frame numbers indicate that this is very slow zoom, taking more than 10 seconds to complete.

4. Tracking Shot - When the camera follows a moving subject, the effect is the tracking shot. The action of tracking is an important portion of a video sequence because it usually shows meaningful transition. It is used in all types of video for different reasons. In Figure 2.11(D), the camera tracks the subject leaving a helicopter (Planet Earth II, WQED). Tracking is quite common in sports video, where cameras follow an athlete of focus or attention during play. Although we do not specifically detect tracking shots, its flow pattern is usually characterized by camera or object motion and subsequently used for summarization.

5. Zoom-in Shot - The zoom-in is used to focus on a particular subject. This procedure is characterized by a narrowing of perspective and visual concentration. It is used in all types of video and is not limited to a fixed location on the screen. The zoom-in shot may be detected through camera motion analysis. An example of this shot is shown in Figure 2.11(E), where the camera zooms into a scientist sitting at a telescope (Planet Earth II, WQED). Using this analysis, we may also detect a zoom-out shot, which is sometimes less important.

2.3.4 Subject Positioning

The position of the subject is an important and simple procedure for conveying a specific theme in video. Two common procedures, which concentrate on positioning, are Viewer Dialogues and Close-ups.

1. Viewer Dialogue - This involves a character on screen talking directly to the viewing audience. This effect was popular in suspense movies prior to the 1950's. It is seldom seen today and is mostly used as a method of comic relief. Broadcast news employs constant anchorperson and viewer dialogue, but not as a special effect. Human face detection is possible in some cases, as discussed in chapter 3. Unfortunately, it is very difficult to distinguish viewer dialogue from other face-forward scenarios in video.

2. Close-up Shots - When an object or person is placed close to the camera, it consumes the majority of the viewing space and serves as the dominant subject in the scene. This effect is used in varying degrees with all types of video. When the subject is a human, detection of the face, position, and size may be possible. If the face is large and positioned with other objects in the background, this is likely a close-up shot.

2.3.5 Angle Shots

The position of the camera is often angled to convey a certain effect. Placing the camera at a particular angle may alter changes in height and viewing space. The two most common forms of angle shots are described below.

1. High and Low Angle Shots - Changes in the height from which an image is shown is a useful procedure for introducing various perspective to a

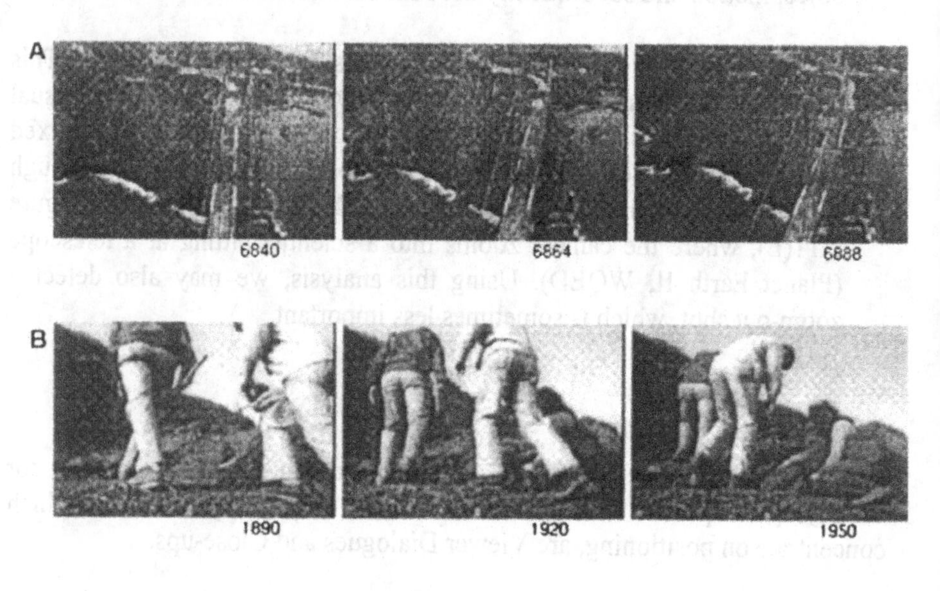

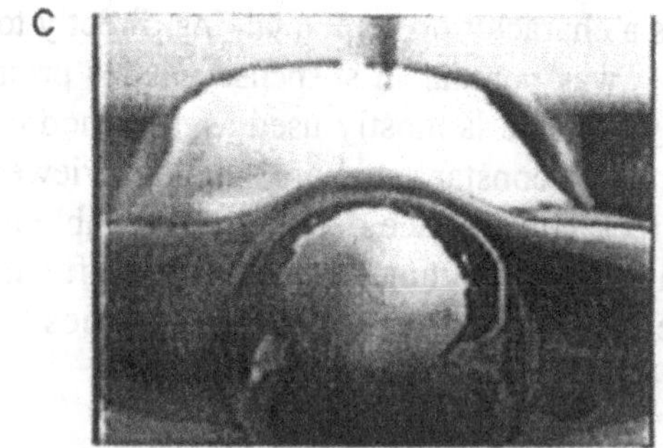

Figure 2.12 Examples of various angle shots: (A) High-Angle Shot (same as Figure 2.11(C)), (B) Low-Angle Shot from the ground upwards, (C) Wide-Angle close-up shot.

scene. For low angle shots, the camera is below the subject shooting upwards. The viewpoint is unusual and can be used to distort the scale of an object in view. The subject is usually the focus with this effect. High angle shots also give an unusual perspective on a scene. In this case, the scenery, not the subject, is usually the focus of the shot. Figure 2.12(A) is an example of a high angle shot over the Golden Gate Bridge in San Francisco, CA (Planet Earth I, WQED). Figure 2.12(B) is an example of a low angle shot with archeologist examining rocks on a mountainside (Planet Earth I, WQED). Angle shots are difficult to detect since they require understanding of camera position and object perspective.

2. Wide-Angle Shot - This procedure is used to create a distorted sense of depth in a scene. A wide-angle camera provides a sharp focus for objects in the foreground and background. The effect of the wide-angle lens may stretch and distort the image, particularly when the subject is close to the camera as seen in figure 2.12(C). This procedure is also quite difficult to analyze. Detection would involve understanding the focus of all objects in a frame and recognizing which objects are distorted with respect to the camera position.

2.3.6 Camera Focus

The detection of changes in camera focus is useful for isolating a region of interest in a video frame. By altering the focal point of the camera, a producer can select which elements will remain in focus and clear for viewer inspection. Several common examples of camera focus effects are illustrated below.

1. Pulling Focus - This procedure occurs when the point of camera focus changes during the course of a scene. Selective focusing can be used as a creative tool to shift emphasis from one part of the scene to another. This effect is quite common in feature-films and sparsely used in documentaries.

   To detect Pulling Focus, we can observe the percentage of screen focus over time for a single scene. If there is a large unfocused region that vanishes over the scene or the point of focus changes, then this is likely the pulling focus effect. There are various techniques in focus detection that make this possible, some of which are described by Shafer's work on camera calibration for stereoscopic analysis [Shafer 1994].

2. Isolating Focus - This procedure is used to show the subject at a distance and with the surrounding foreground unfocused. By reducing the depth of field, or the distance from the camera within which the subject is sharply focused, the area surrounding the subject is blurred and unrecognizable. A wide aperture is used to achieve this effect. Isolating focus scenes appear in some degree in all types of video, but they are most common in feature-films. Detection of this effect is possible if areas of sharp focus can be discriminated from blurred areas.

3. Shallow Focus - The shallow focus shot is created by applying a long focal length to reduce the depth of field. This places the background out of focus and the subject in focus and thus highlighted. This effect is similar in appearance to the isolating focus shot except that the subject is in the foreground. It is used casually in all types of video, especially when the subject is shown against a distant background.

2.3.7 Lighting and Mattes

Lighting effects are often used to convey a certain mood in video [Millerson 1991]. Adjusting the amount of light on a subject or background is a simple technique for creating the emotional overtone in a scene. Several lighting effects are discussed in the sections below.

1. Low Lighting - A low lighting effect is produced when a directional light source is positioned at an angle to the subject. This will emphasize any textural details and creates a high contrast between highlights and shadows. Low lighting is often used to convey tension or mystery, and is primarily used in feature-films. It is difficult to detect since the light position is unknown and the properties of the image are similar to other effects.

2. Dramatic Lighting - In most video, the lighting is arranged to balance the areas within a scene so that the subject is evenly lit with no shadows. Varying the degree of light and position to affect contrasting areas of light and darkness creates dramatic lighting. It is used to highlight a subject or scene and convey a certain mood. This procedure is also difficult to detect because its image statistics are similar to other effects. Potential detection is further complicated by inconsistent lighting patterns used by different producers.

3. Silhouettes - The silhouette shot is an effect in which all light sources originate from the background. This causes characters to appear

mysterious and actions more dramatic. Usually there is little visible detail so perceptual recognition of the subject is less important than the mood of the scene. The subject is often a human face in profile. Silhouettes appear most often in feature-films and seldom in documentaries. Detection of silhouettes is difficult since their image statistics are similar to other types of image lighting effects. The isolating focus effect from the prior section uses a silhouette to show a viewing subject.

4. Matte Effects - When a matte is placed in front of the camera, the illusion is created that the viewer is watching a scene through that matte. Matte effects convey the visual impression of a viewing device, such as telescopes, binoculars, or masks. These effects appear most often in feature-films, and it is common to use the telescope or binocular matte in animal and exploratory documentaries. Examples of a telescope and highlight matte are shown in Figure 2.13. This effect is quite easy to detect when the matte pattern is known. The detection of these mattes is further simplified when a monotone black or white matte pattern is used.

Figure 2.13 (Left) Telescope Matte for artistic enhancement, (Right) Highlight Matte from an infrared camera tracking a human subject across a parking lot. The matte is slightly transparent to display the background and location.

2.3.8 Grayscale Video

Until 1964, most video was limited to grayscale imaging. When grayscale video appears in modern color-possible media, it usually provides historical perspective and is perceived as material from the past. This is certainly the case with documentaries, news, and sports footage. In feature-films, grayscale video may imply older footage or an aged effect. An example of combined grayscale and color video is shown in Figure 2.14 (The History of Plastics - Documentary, WQED, Pittsburgh, PA).

A histogram of a grayscale image should have approximately equal values of **RGB** for a given pixel, making detection possible for a stream of images in video. The analysis and detection of grayscale video is discussed in chapter 3.

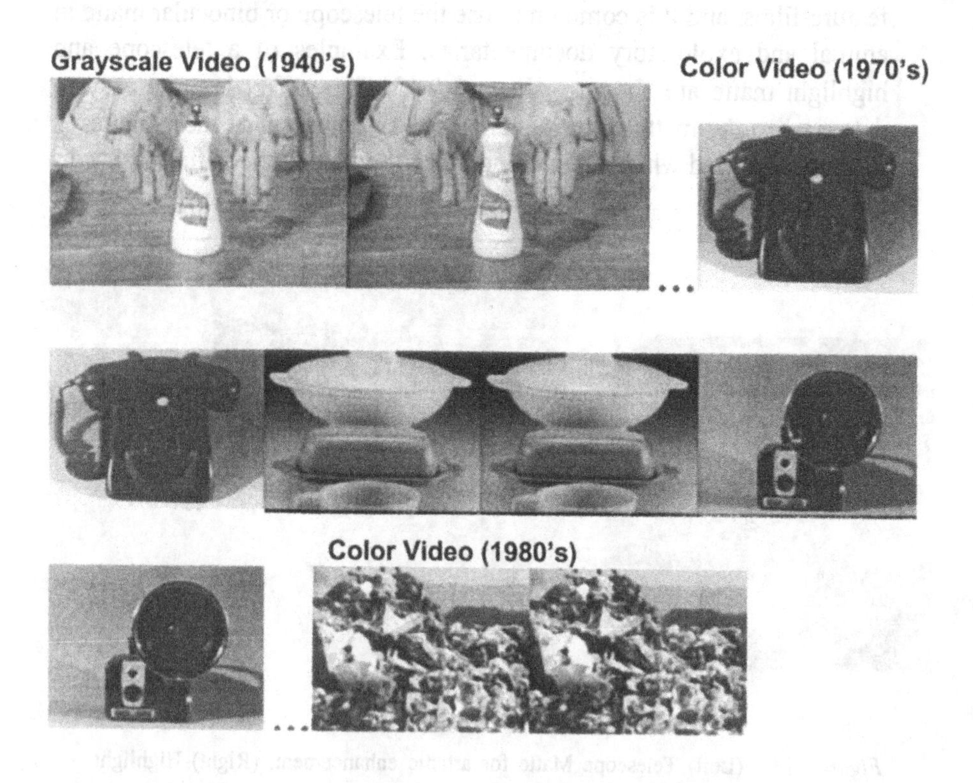

Figure 2.14 Grayscale video inserted in a documentary to indicate past or historical content. The grayscale video is from a commercial produced in the 1940's. It is used to start a time sequence of videos on the history of plastics. The subsequent color footage contains video from the 1970's and 1980's

2.3.9 Audio Effects

Audio effects, in addition to normal speech, music, noise or silence, are very important for conveying the theme in a video [Zhang 1999]. Some of the more common audio effects are discussed below.

1. Special Sound Effects - A producer can use special sound effects to convey concepts too difficult to create visually. Sound effects can occur naturally in documentaries, news footage and sports. In some video and most feature-films, a sound effect may be added synthetically. Examples of these effects include:

   - Mechanical sounds such as robots, machinery, engines, etc.,
   - Weather related sounds such as wind, rain, etc.,
   - Animal sounds

   Although a particular sound may be possible to model, sound effect recognition, in general, is quite difficult.

2. Loud Sound Effects - In addition to special sound effects, there are often very simple sounds that heighten the mood of a scene through audio amplification, such as:

   - Explosions
   - Crashes or collisions
   - Gunfire

   Recognition and discrimination of loud sound effects, in general, is quite difficult. The pattern may change dramatically between segments. An approximate solution for detection is to isolate regions of increased audio amplitude. Although this will not distinguish an explosion from a collision, for example, we may gain understanding into the type of action taking place. Loud sound effects are added synthetically in many feature-films or they occur naturally in other types of video and may provide important insight for prioritization.

3. Screams - A common sound that may be detected through amplitude gains and localized frequencies is the human scream. The act of screaming, yelling or some other loud human sound is often used in feature-films to heighten tension and suspense. This effect appears naturally and with less frequency in other types of video.

4. Theme Music - A segment will often contain one or more recognizable sound tracks called theme music. This is most common in feature-films, although other types of video may also use some form of theme music. The movie "Star Wars" contains a widely recognizable sound track where a particular music piece is used to convey a known mood or to indicate transition. Documentaries usually do not contain separate audio tracks for merchandising, but they do contain transitional music and some theme music.

5. Audio Discrimination - Detection of simple sound patterns is more plausible when the music is not combined with human speech [Rabiner 1976]. With advances in speech analysis, it is possible to automatically distinguish music from speech in certain audio signals. This is more difficult when speech and music are combined in the same track, such as opening monologues, action movies and most songs. Kedem introduced the method of Zero-Crossing for audio discrimination that is still widely used today [Kedem 1986]. Saunders explored real-time applications of audio discrimination in broadcast speech [Saunders 1996]. Other researchers have proposed methods by frequency analysis of compressed audio streams [Jarina 2001].

2.3.10 Word Selection

Language effects, in addition to standard human dialogue, are useful in evaluating the importance of words in the audio track. Certain words may be recognized to convey a particular theme by themselves without any additional semantic understanding of a phrase or paragraph. Unlike audio and image understanding, where accuracy is dependent on imperfect audio and image capture technologies, language understanding is based on a text transcript. In most cases, a near perfect digital transcript is available, so detection of a particular word is highly accurate. Certain words that appear in the transcript are more important than others and should be included in a video summary. Some of the more common language selection procedures are discussed below.

1. Low Frequency Words – A word that appears less frequently in a certain standard corpus is a significant indicator of the video content when it appears in a video. A word like "Extinct" may never appear in most news broadcasts, but would be quite common in documentaries on archeology or animal species.

2. Slang Words - The use of "Slang" words is quite common in feature-films that often attempt to mimic natural human speech. Detection of certain slang words may help identify areas of tension in a video, if that particular slang word is seldom used in the ordinary video.

3. Conjunction Words - Sentences with certain conjunction words imply a change in semantic direction. Words such as "therefore", "but", "nevertheless", and "although", imply a change in the thought process.

4. Question Words - A question statement may provide added insight by introducing some problem or issue related to the topic. Sentences with questions words are recognizable since the domain is somewhat limited. A question statement will usually contain one of the following words: "How", "When", "Where", "What", or "Why". Question statements are particularly useful in documentaries since a segment will often start by asking a relevant question.

2.4 Visualization Systems in Use Today

Video summarization, segmentation and other visualization interfaces are used in present day video presentations. The concept of browsing and viewing shortened video is pervasive in many video applications.

Video Summaries

Video summaries have great potential for use in digital video libraries, as well as other mediums that use video. A number of summary representations similar to the video skims research from Carnegie Mellon University [Smith 1997] are used in broadcast television. Examples of these representations include the following:

1. News Summaries - Many news stations provide a short summary of regional, state, or world news. An example of a news summary is the "World in a Minute" from WPXI, Channel 11 News, Pittsburgh, Pennsylvania. As the title indicates, this one-minute segment displays several world news stories, each usually 8 seconds in duration. The purpose of the news summary is to display the most important news from around the world in a short time.

2. Sports Highlights - Sports highlights are common in local and world news. An example of a sports highlight segment is the "Sports Machine" with Al Michaels, CBS. Even though this show has commentary, a large

portion is devoted to specific sporting events. A segment shows several isolated portions of a sporting event with commentary as audio. The purpose of the highlight video is to convey points of interest in a short amount of time.

3. Recorded Sports Broadcasts – A sporting event that is rebroadcast at some later time is often edited to remove ambiguous content or to shorten the duration. In Japan, baseball games that are rebroadcast need to fit into a one-hour time slot. The plays are parsed manually to create a shorter version of the original game. College football in the U.S.A. is often rebroadcast in the Pacific Time Zone for viewers who miss early games that take place in the Eastern Time Zone. For these broadcasts, highlights are aired that usually include scores, long plays and turnovers.

4. Movie Trailers and Previews - Short video trailers and previews are produced to attract potential customers for feature-films. Conveying the content truthfully is not the primary motivation. The selection of segments is based more on producer preferences, which includes exciting and climactic video.

5. Introductory segments - Most documentaries include a short video abstract prior to the full segment. They contain image and some audio portions from the full segment. The purpose of the abstract is to convey the overall content of the later segments.

Browsing

A number of commercial browsing systems exist today, both as software and hardware devices. They include:

- DVD – Chapter skipping and digital Fast Forward and Reverse
- Analog Fast Forward – A noisy medium that distorts
- Accelerated Playback – Pitch control technology enables the acceleration and deceleration of audio or video playback without annoying high frequency artifacts [Nakamura 1996] and [Seiyama 2001]. These systems exist primarily in software.
- Roll Bar – Cylindrical rolling tool in many analog videocassette recorders and DVD systems. Provides greater user control to video access points.

Segmentation in Media

- Commercial Skipping – Detects black frames or skips ahead by some fixed increment (usually 30 seconds).
- GoTuit Inc. – Manual segmentation and annotation of popular television sitcoms and news.
- Mate Inc. – Developed a system that searches for content on multiple channels based on a user-specified profile or query. Search takes place during a live broadcast and alerts the user when content of interest is found on a different channel.
- TiVo Inc. – System to search for content based on a user-specified profile and record for later viewing.

Visualization from Characterization in Media

Many systems that exist today require manual intervention to first create characterization data for content description and then to visualize video based on it. Examples include:

- DVD Thumbnails - Most commercial DVD contain thumbnails that mark chapters in a video.
- Digital Camera Software - Digital cameras are sold with video editing software capable of detecting shot changes and storing thumbnails for previewing.
- Audio and Video Editing - Current editing systems provide temporal characterization data in addition to the conventional image and audio timeline. This characterization data usually includes shot changes and video effects such as audio enhancements, fades, dissolves, etc.
- Video Logging - Similar to editing, logging is the procedure for adding annotation or characterization information to a video. Many commercial and academic systems provide automated and human-assisted forms of logging.
- Video Descriptive Services (VDS) - Some Public Broadcast Stations (PBS) provide this service for the blind. As a video is displayed, a narrator describes the visual content. Figure 2.15 is an example of VDS shown with closed captions.

2.5 Higher Level Interpretation for Summarization

This book emphasizes the importance of characterization for summarization and segmentation. Semantic video understanding may seem intuitively straightforward, but the resultant summary or segmentation is different from one video editor to another [Christel 1997]. A professionally trained editor can recognize important themes, such as introduction and climax. For example, most documentary video is structured with an opening introduction, followed by supportive examples. Figure 2.16 shows an example of this with a documentary video from WQED, Pittsburgh, PA. The video starts with an introductory segment and introduces a related human subject. The main topic of "Artificial Skin" is then established and presented with expert interviews. Supportive examples and a closing segment follow this portion on the original human subject. Figure 2.17 shows the format of broadcast news, which includes an introduction, narration, topical examples and closing remarks.

Summary Creation from Video Characterization

The characterization data is only useful if a pattern or rule is established to prioritize image and audio information. To study the potential for creating such rules for image and audio selection, an experiment was conducted in 1994 with manually created skims for one-hour documentary videos at various compaction levels. Producers and technicians in the Carnegie Mellon University Drama Department provided suggestions for video production procedures that could be detected automatically. Distinct patterns emerged from common video effects; and thus, the potential for video skims was realized. The study revealed that while perfect skimming requires semantic understanding of the entire video, certain parts of the image and audio selection process could be automated with only partial semantic understanding that current image, audio, and language understanding technologies can provide. The following chapters address the issues related to automatic detection of audio and image characteristics and the parameters involved in creating visualizations for summarization and segmentation.

2.6 Conclusions

This chapter has introduced video terminology at the research and production level. It is important to note the standards and formats used by professional producers when developing systems and applications for video characterization. In chapter 3, we use these standards as a basis of developing automated technologies for feature detection in image, audio and language.

Audio: An unlikely setting in which to begin the exploration of our solar system

VDS: **Two scientists walk away from a helicopter that has landed on ice**

Audio: Meteorites are trapped in a huge conveyor belt of ice and carried to the heart of the continent to the sea

VDS: **Wind moves violently across the snowy surface as the helicopter departs**

Audio: By analyzing this ancient material ...

Figure 2.15 Graphical depiction of closed captions and Video Descriptive Services (VDS). Without VDS, a blind viewer may not realize the footage is shot in a cold environment. The majority of the dialog or closed captions describes the actual subject matter, "Meteorites contribution to understanding our solar system".

Figure 2.16 Structured Documentary Format – In this documentary, an introduction is made by first showing a climatic event. The subject is then introduced in conjunction with follow-up commentary. In many cases the subject of an interview is not a professional speaker, and speech recognition and audio analysis is quite difficult. We discuss the limitations of audio analysis in conventional video in section 3.3. The remainder of the segment toggles between subjects and examples of a new surgical procedure for synthetic skin – which is the subject of the documentary. The closing scene is a transition to another topic, which contains soft music and a departing tracking shot.

Introduction (Transition from previous story to K'nex Toy Story)

Narration with Camera Zoom

Narration with Example

Narration with Example

Correspondent Narration and Closing

Figure 2.17 Broadcast news structure: Introduction (row 1), Local Correspondent Narration (row 1), Examples with transitional camera zoom-in (rows 2, 3 and 4), Correspondent Narration and Closing Remarks (row 5).

2.7 Bibliography

Aggarwal, J. K. and Cai Q., "Human Motion Analysis: A Review" *Computer Vision and Image Understanding: CVIU* 1999.

Bordwell, D., Thompson, K., "Film Art: An Introduction," 4th Ed., McGraw Hill, Englewood Cliffs, NJ, 1993.

Boreczky, J.S., and Rowe, L. A. "Comparison of Video Shot Boundary Detection Techniques". *In Storage and Retrieval for Still Image and Video Databases IV, Proc. SPIE* 2664, pp. 170-179, Jan. 1996.

Chang, P., Han, M., Gong, Y., "Extraction Highlights from Baseball Game Video with Hidden Markov Models", *International Conference on Image Processing, ICIP* 2002.

Christel, M., Winkler, D., & Taylor, C., "Multimedia Abstractions for a Digital Video Library". *Proceedings of the 2nd ACM International Conference on Digital Libraries*, R. Allen & E. Rasmussen, eds. (Philadelphia, PA, July 1997), pp. 21-29.

Christel, M.G., Winkler, D.B., & Taylor, C.R. "Improving Access to a Digital Video Library". *Human-Computer Interaction INTERACT '97*: IFIP TC13 International Conference on Human-Computer Interaction, 14th-18th July 1997, Sydney, Australia, S. Howard, J. Hammond & G. Lindgaard, Eds. London: Chapman & Hall, 1997, pp. 524-531.

Collins, R., Amidi, O., and Kanade, T., "An Active Camera System for Acquiring Multi-view Video", *Proceedings of the 2002 International Conference on Image Processing* (ICIP '02), September, 2002.

Ekin, A., and Tekalp, A. M., "Automatic Soccer Video Analysis and Summarization", *Electronic Imaging: Science and Technology: Storage and Retrieval for Image and Video Databases IV*, IS&T/SPI03, Jan. 2003, CA.

Ekin, A., and Tekalp, A. M., "A Generic Event Model and Sports Video Processing for Summarization and Model-Based Search", in the Handbook of Video Databases (ed. Borko Furht and Oge Marques), CRC Press, 2003.

Escamilla R, Francisco A, Fleisig G, Barrentine S, Welch C, Kayes A, Speer K, Andrews J, "A Three-dimensional Biomechanical Analysis of Sumo and Conventional Style Deadlifts", Division of Orthopedic Surgery, Duke University Medical Center Medicine and Science in Sports and Exercise 2000 Jul;32(7):1265-75

Gargi, U., Kasturi, R., Strayer. S.H., "Performance Characterization of Video-Shot-Change Detection Methods". *IEEE Transaction on Circuits and Systems for Video Technology*, Vol. 10, No. 1, February 2000.

Hampapur, A., Jain, R., and Weymouth, T. "Production Model Based Digital Video Segmentation", *Multimedia Tools and Applications*, 1 (March 1995), 9-46.

Jarina R, Murphy N, O'Connor N and Marlow S. "Speech-Music Discrimination from MPEG-1 Bitstream", SSIP'01 - *WSES International Conference on Speech, Signal and Image Processing*. Malta, 1-6 September 2001.

Jones, M., J., Rehg, J., "Statistical Color Models with Application to Skin Detection", *Computer Vision and Pattern Recognition -CVPR*, Ft. Collins, CO, Pages 274-280, June 1999.

Ju, X. S., Black, M.J., Minnerman, S., and Kimber, D., "Summarization of Videotaped Presentations: Automatic Analysis of Motion and Gesture", *IEEE Transactions on Circuits and Systems for Video Technology*, Vol. 8, No. 5, September 1998.

Kedem, B., "Spectral Analysis and Discrimination by Zero-Crossings", *Proceedings of the IEEE, vol 74 no 11, November 1986.*

Kobla, V., DeMenthon, D., Doermann, D., "Detection of Slow-motion Replay Sequences for Identifying Sports Videos", *IEEE 3rd Workshop* on *Multimedia Signal Processing*, 1999, Page(s): 135 –140.

Li B., and Sezan, I.M., "Event Detection and Summarization in Sports Video," *Proc. IEEE Workshop on Content-Based Access to Video and Image Libraries*, IEEE CS Press, Los Alamitos, Calif., 2001.

Lienhart, R., "Reliable Transition Detection In Videos: A Survey and Practitioner's Guide". MRL technical report; *International Journal of Image and Graphics* (IJIG), August 2001.

Metoyer, R.A. and Hodgins, J.K., "Animating Athletic Motion Planning By Example," *Proceedings of Graphics Interface 2000*, Montreal, Quebec, Canada, May 15-17, pp. 61-68.

Millerson, G. "Technique of Television Production" 12th Ed.. Butterworth-Heinemann Publishers 1990.

Millerson, G. "Technique for Lighting of Television and Film" 3rd Ed. Butterworth-Heinemann Publishers 1991.

Millerson, G. "Video Camera Techniques – A New Media Manual" 2nd Ed.. Butterworth-Heinemann Publishers 1994.

Millerson, G. “Video Production Handbook” 3rd Ed. Butterworth-Heinemann Publishers 2001.

Nakamura, A., Seiyama, N., Imai, A., Takagi, T., and Miyasaka, E., "A new approach to compensate degeneration of speech intelligibility for elderly listeners," *IEEE Transactions on Broadcasting.*, vol.42, no.3, Sept. 1996.

Pan, H., Li, B. and Sezan, M. I., “Detection of slow-motion replay segments in sports video for highlights generation,” *Proceedings of the IEEE International Conference on Acoustic, Speech and Signal Processing*, ICASSP 2002.

Pingali, G.S., Opalach, A., and Y. Jean, Y., “Real-time Ball Tracking for Innovative Tennis Broadcasts”. *Proceedings of the International Conference on Pattern Recognition* (ICPR), pp. 152-156, Vol. 4, Barcelona, September 2000.

Pryluck, C., Teddlie, C., Sands, R., “Meaning in Film/Video: Order, Time and Ambiguity,” Journal *of Broadcasting* 26, pp. 685-695, 1982.

Rabiner, L. R., Cheng, M. J., Rosenberg, A. E. and McGonegal, C. A. "A comparative performance study of several pitch detection algorithms". *IEEE Transactions on Acoustics Speech and Signal Process.*, Volume ASSP-24, pp 399-417, Oct. 1976.

Rees, D., Agbinya, J.I., Stone, N., Chen, F., “CLICK-IT: Interactive Television Highlighter for Sports Action Replay”, *Proceedings, Fourteenth International Conference on Pattern Recognition*, Volume: 2 , Pages: 1484 –1487, 1998.

Sato, T. Kanade, T., Hughes, E. Smith, M. “Video OCR for Digital News Archive,” *IEEE International Workshop on Content-based Access of Image and Video Databases*, ICCV, Bombay, India, January 1998.

Saunders, J., Real-Time "Discrimination of Broadcast Speech/Music", *ICASSP* 96, vol. 2 (pp 993-996).

Seiyama, N., Imai, A., Mishima, Takagi T., and Miyasaka, E., "Development of high-quality real-time speech rate conversion system," *IEICE* vol.J84-D-2 No6. Pp.918-926 Jun. 2001.

Smallman, K., “Creative Film-Making”, 1st Ed., MacMillan Publishers, New York, 1970.

Smith, M., Kanade, T. “Video Skimming and Characterization through the Combination of Image and Language Understanding Techniques”. *Computer Vision and Pattern Recognition*. San Juan, PR, June 1997.

Sudhir, G., Lee, J.C.M., Jain, A.K., “Automatic classification of tennis video for high-level content-based retrieval”, *Proceedings IEEE International*

Workshop on Content-Based Access of Image and Video Database, 1998, Page(s): 81 –90.

Stolcke A., et al, "Dialogue Act Modeling for Automatic Tagging and Recognition of Conversation Speech", *Computational Linguistics 2000*, Vol. 26, Number 3.

Wactlar, H.D., Kanade, T., Smith, M.A., and Stevens, S.M. "Intelligent Access to Digital Video: Informedia Project," *IEEE Computer*, 29, May 1996, 46-52.

Wactlar, H., Hauptmann, A., Smith, M.A., Pendyala, K., Garlington, D. "Automated Video Indexing of Very Large Video Databases," *SMPTE Journal*, August, 1997.

Xiong Y., and Shafer S., "Moment and Hypergeometric Filters for High Precision Computation of Focus, Stereo and Optical Flow", *International Journal of Computer Vision*, Vol. 22, No. 1, February, 1997, pp. 25-59.

Yong Rui; Anoop Gupta; Alex Acero; Automatically Extracting Highlights for TV Baseball Programs, *Proceedings of the ACM International Conference on Multimedia*, Oct. 2000, Los Angeles USA, Pages 105 –115.

Yow, D., Yeo, B.L., Yeung, M., and Liu, G., "Analysis and Presentation of Soccer Highlights from Digital Video" *ACCV Proceedings*, 1995, Singapore, Dec. 5-8, 1995

Zhang, H.J., Tan, S., Smoliar, S., and Yihong, G. "Video Parsing, Retrieval and Browsing: An Integrated and Content-Based Solution," *Proceedings of the ACM International Conference on Multimedia*, San Francisco, CA, November 1995.

Zhang T.and C.-C. J. Kuo, "Hierarchical classification of audio data for archiving and retrieving", *Proc. ICASSP'99*, Vol. 6, Phoenix, Mar. 1999.

Zhou, W., Vellaikal, A., et al., "Rule-based Video Classification System for Basketball Video Indexing", *Workshop - Proceedings of the ACM International Conference on Multimedia*, 2000.

Chapter 3

Multimodal Video Characterization

Video characterization is the extraction of important information to create a synopsis of the original video. Digital media allows video understanding methods to analyze video in ways not possible with traditional analog media. For image understanding, characterization includes segmentation of video into shots or scenes, detection of important objects and identification of the structural motion of a scene. For audio and language understanding, it entails performing syntactic, semantic and statistical analysis of associated sound and text in order to, for example, locate human speech, spot the most significant words, and create summary sentences.

Section 3.1 discusses representations for video characterization, in particular, temporal characterization displays. Section 3.2 defines video features. Sections 3.3 and 3.4 cover various automated methods for audio and image feature detection. Section 3.5, with contributions by Tsuhan Chen, Department of Electrical and Computer Engineering, Carnegie Mellon University, is devoted to compressed domain features.

3.1 Video Characterization Representations

Video characterization represents the temporal structure of video content over time. The representation must establish a correlation between video metadata and video content.

Several visual representations have been proposed in video analysis. They display imagery as thumbnails, audio by some acoustic sound measure, and content by the features associated with the video.

A simple form of temporal characterization is a thumbnail display. Figure 3.1a is an example of a thumbnail display of several images from a video. The images provide an overview in less time than watching the original video. Thumbnails have been used in video editing systems for many years and are now being used in video libraries and other asset management systems.

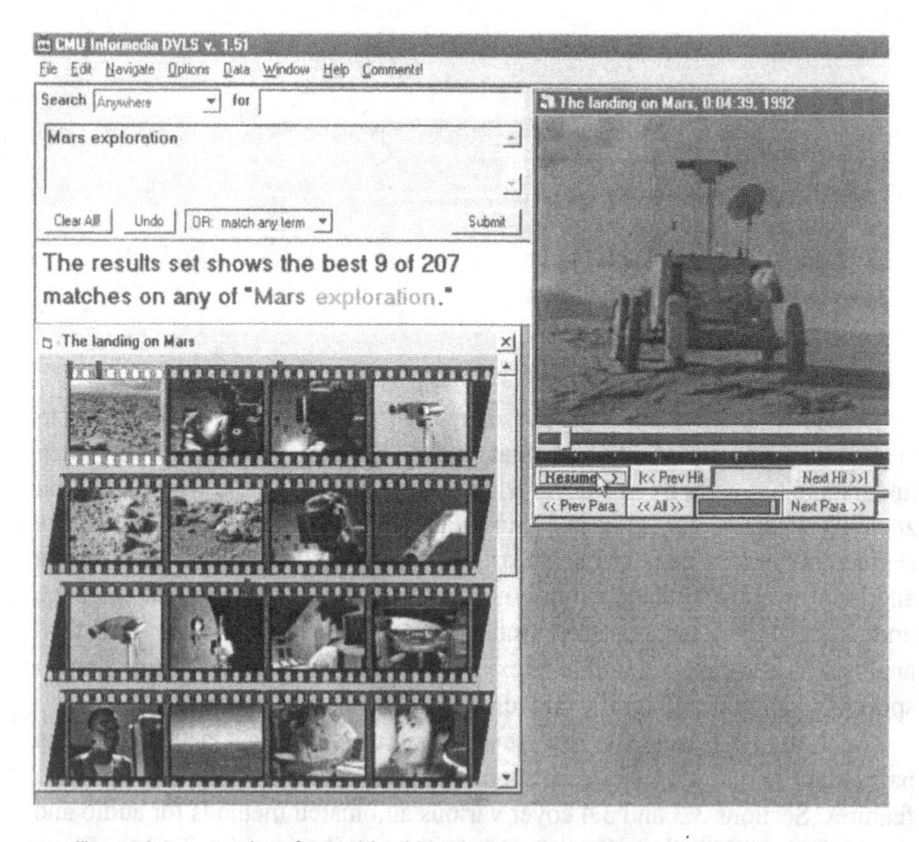

Figure 3.1a An interface with video thumbnails. The thumbnails provide a temporal view of the video sequence without playback.

Figure 3.1b illustrates a Temporal Characterization Graph (TCG), used by Smith and Kanade [Smith 1997], of a segment taken from the "Destruction of Species" documentary, WQED Pittsburgh. Video is segmented into shots, and camera motion is detected along with significant objects (faces and text); horizontal bars indicate regions with positive results. Word relevance is evaluated in the transcript.

The TCG provides a view of the video content and its metadata by aligning them along the time axis. It is similar to traditional audio and video editing displays, but with descriptive content information extracted from the video. Collectively this content information is the foundation for interpreting video. For example, the vertical box in Figure 3.1b shows the portion that contains a camera pan, a human face, and spoken dialog.

3.2 Video Features

A feature is a descriptive parameter that is extracted from audio, images, text or video. Features may be used to interpret audio or visual content, or as a measure for applications such as summarization, segmentation or similarity retrieval in image and video databases.

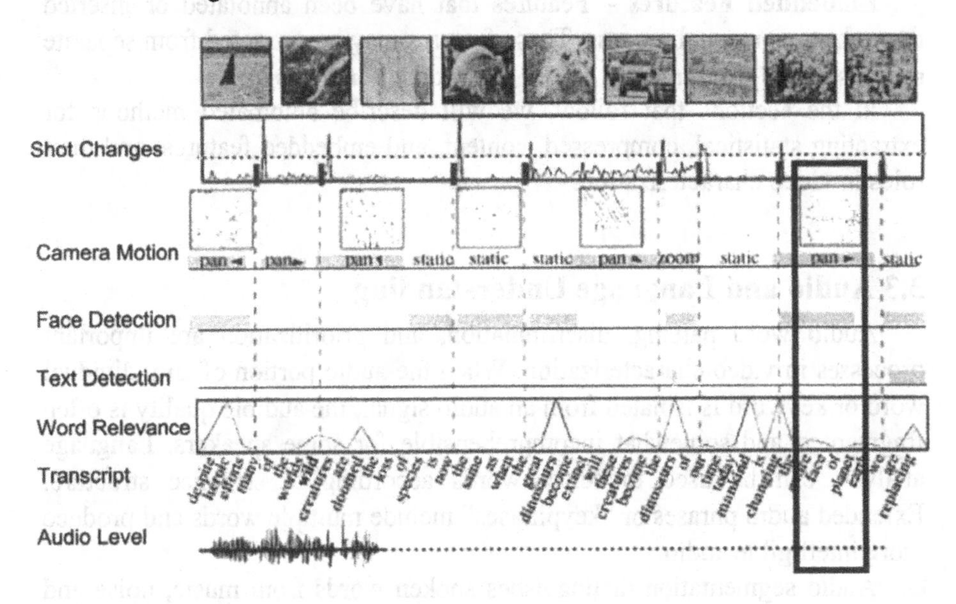

Figure 3.1b Temporal Characterization Graph - Methods for Video Characterization: video is segmented into shots, and camera motion is detected along with significant objects (faces and text). Horizontal bars indicate regions with positive results. Word relevance is evaluated in the transcript. The vertical box to the right shows the portion of the video that contains a camera pan, a human face, and spoken dialog.

Statistical Features – Low-level features that are extracted from an audio or image sequence in terms of certain statistical measures without regard to content. These include parameters derived using such algorithms as image difference, color histogram, word count, and audio volume. Section 3.3 describes statistical features in audio and language understanding. Section 3.4 describes those in image understanding, in particular, color, texture and difference features.

Compressed Domain Features – Features extracted directly from a compressed image or video stream without regard to content. The method has an advantage of efficient processing. Use of the DCT (Discrete Cosine Transform) coefficient is one example. Section 3.5 expounds on these topics with additional discussion on the MPEG-7 standard.

Content Features – Features derived for describing the actual content in an audio, image or video stream. For example, optical flow itself is a statistical feature, but the interpretation of optical flow as camera motion is a content feature. Section 3.4 describes methods for interpreting motion, on-screen text and certain objects in video by means of computer vision techniques.

Embedded Features – Features that have been annotated or inserted through some manual process. These features may be extracted from separate media such as programming guides and production notes.

In the sections that follow, we will describe automated methods for extracting statistical, compressed, content, and embedded features, and their roles in video characterization.

3.3 Audio and Language Understanding

Audio word parsing, discrimination, and prioritization are important processes in video characterization. When the audio portion of an individual word or keyword is isolated from an audio signal, the audible quality is often fragmented and somewhat incomprehensible for some speakers. Language analysis can be used to select words according to sentence structure. Extended audio phrases or "keyphrases" include multiple words and produce more intelligible audio.

Audio segmentation distinguishes spoken words from music, noise and silence. Further analysis is necessary to align and translate these words into text. Word parsing is made on a frame-by-frame basis, so it is important to achieve the highest possible accuracy. At a sampling rate of 8 KHz, one video frame corresponds to 267 samples of audio.

3.3.1 Speech Transcription

Automated speech recognition is used to transcribe audio content and select representative words. The Carnegie Mellon University's Sphinx-II speech recognizer [Hwang 1994] was used for the experiments described below in this section.

In many cases, the audio has inconsistent audible speech and generates errorful transcripts. The accuracy levels obtained in speech recognition experiments for video audio are listed in Table 3.1. The data was compiled from experiments with documentaries and broadcast news in a variety of acoustic environments [Hauptmann 1995].

Recognition accuracy improves when closed-captions are available and used to align and supplement erroneous recognition. Captions usually occur in broadcast material, such as sitcoms, sports, and news. Documentaries and movies may not necessarily contain captions. Closed-captions have become more common in video material throughout the United States since 1985 and most televisions provide standard caption display.

Closed-captions provide timing offset and ASCII text for groups of words in the transcript [Seidenberg 1981]. An example is shown in Figure 3.2. The timing offset approximates the location of the captioned words in the audio track. The alignment of these words with the actual spoken audio is accurate to 1 or 2 seconds with recorded or edited video. Live footage may exhibit a delay of up to 20 seconds or more.

Once a transcript is generated, semantic parsing isolates meaningful audio, such as noun and verb phrases [Sleator 1993]. Figure 3.4 lists examples of noun phrases detected in a transcript.

3.3.2 Language Characterization

For documentaries, a digital ASCII version of the transcript is usually attached with an analog version of the video. Keywords and keyphrases are identified through language analysis. A well-known technique is TF-IDF (Term Frequency Inverse Document Frequency), which measures relative importance of words in text and video documents [Salton 1983], [TREC 1993], [Mauldin 1991], and [Siegler 1999]. Specifically, the TF-IDF of a word is the ratio of the frequency f_s of a word in a given scene to the frequency f_c of its appearance in a standard corpus.

$$TF-IDF = \frac{f_s}{f_c}$$

035012.7966319444 >>> GOOD EVENING.

035012.7966435185 I'M LAURIE BISHOP.

035012.7966782407 BLACK TUESDAY COULD BE LOOMING

035012.7966898148 FOR THE FEDERAL GOVERNMENT.

035012.7967129630 THE CLINTON ADMINISTRATION IS

035012.7967361111 WARNING THAT UNLESS CONGRESS

035012.7967476852 PASSES A STOPGAP BILL BY NEXT

035012.7967824074 TUESDAY, THE FEDERAL GOVERNMENT

035012.7967939815 WILL BE FORCED TO CLOSE

035012.7968055556 ITS DOORS.

035012.7968287037 WOLF BLITZER EXPLAINS WHAT

035012.7968402778 A SHUTDOWN COULD MEAN

035012.7968518519 FOR AMERICANS.

035012.7968865741 >> Reporter: THE WHITE HOUSE

035012.7968981481 SAYS 800,000 FEDERAL WORKERS

035012.7969097222 AROUND THE COUNTRY WILL HAVE

035012.7969328704 TO BE TEMPORARILY LAID OFF NEXT

035012.7969560185 TUESDAY IF THE REPUBLICAN-LED

035012.7969791667 CONGRESS DOESN'T PASS ACCEPTABLE

035012.7970023148 SHORT-TERM LEGISLATION TO AVOID

035012.7970254630 A GOVERNMENT SHUTDOWN.

Figure 3.2 Sample Closed-Caption text from CNN Headline News (CNHAE.cc). The numbers are millisecond offsets between phrases.

Table 3.1 Speech Recognition Results [Hauptmann 1995]

Type of Speech Data	Word Error Rate = Insertion+Deletion+Substitution
1. Speech benchmark evaluations	~ 8% - 12%
2. Speech recorded in speech lab	~ 10% - 17 %
3. Narrator recorded in TV studio	~ 20 %
4. C-Span	~ 40 %
5. Dialog in documentary video	~ 50% - 65 %
6. Evening News	~ 65 %
7. Complete 1-hour documentary	~ 75%
8. Commercials	~ 85%

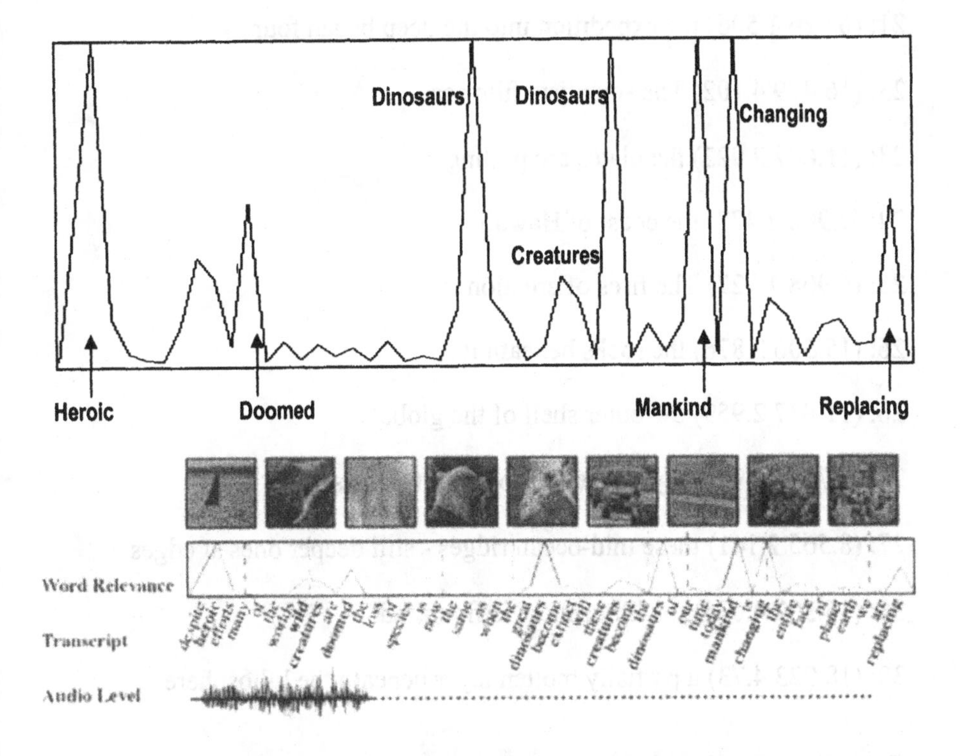

Figure 3.3 Graph of TF-IDF weights (normalized to maximum value) of keywords extracted from a documentary on animal extinction. In this example, the word "dinosaur" has a high TF-IDF score. The word "dinosaur" is used infrequently in daily speech (New York Times text corpus), but it appears 5 times in a 10 minute segment from the documentary.

Words that appear often in a particular segment but appear relatively infrequently in the standard corpus receive the highest TF-IDF weights. An example of TF-IDF weights and the results of keyword selection are shown in Figure 3.3.

Punctuation in the transcript is used to identify sentence boundaries. With sentence boundaries identified, we can parse smaller important regions, such as, noun phrases and conjunction phrases. The link grammar parser [Sleator 1993] was used to parse noun phrases. Sample outputs are shown in

PLE01A

16: (8.562 2.140) the sea floor grows progressively older

20: (10.814 2.163) Alvin will give scientists a first-hand look

20: (4.060 1.015) a first-hand look at a site

21: (6.026 1.506) the expedition into the deep began four

23: (16.409 4.102) The scientists film sea

23: (11.687 2.922) the plates are pulling

24: (9.906 2.477) the coast of Hawaii

25: (6.908 1.727) The fires of creation

26: (15.506 3.876) the rocks beneath it

26: (11.837 2.959) the outer shell of the globe

26: (6.793 1.698) they thicken to some sixty miles

27: (8.565 2.141) these mid-ocean ridges - still deeper ones at edges

28: (16.302 4.075) the lithosphere is fractured

30: (18.923 4.73) a partially molten layer beneath the lithosphere

Figure 3.4a Sample Link Grammar Parser Output and the associative TF-IDF weights (PLE01A). The first number is a modified offset corresponding to frame numbers (30 fps). The first value in parenthesis is the total TF-IDF score for the noun phrase. The second value in parenthesis is the maximum TF-IDF score from the words in the phrase.

Figures 3.4a and 3.4b. With audio alignment from speech recognition, we can extract the part of audio corresponding to the words/phrases with the highest TF-IDF values.

We can also look for conjunction phrases, "slang" words and question words to alter the frequency weighting for the TF-IDF analysis as discussed in section 2.3.10. The TF-IDF values for conjunction and questions words are increased by a factor of 2 to 5 times their original value during keyword detection.

INV15B

17: (13.878 2.776) burke decided to do something

18: (13.147 3.287) the burn is very deep **and**

18: (12.307 2.461) fifteen years is that burn patients

18: (1.684 0.421) be operated on immediately after their injury

22: (13.789 2.758) You have to close the wound

24: (1.145 0.286) working with scientists at m

28: (10.759 1.793) the material is constructed in two layers

30: (7.827 1.957) To give it the look feel **and**

31: (5.385 1.346) The genius of the design is the concept

33: (11.237 1.605) the patient 's body can regenerate its own

33: (3.612 0.903) It serves as a substitute

34: (15.206 3.801) The dermis is freeze-dried

36: (5.262 1.315) A silicone mixture provides the outer layer or epidermis the barrier between the body **and**

38: (3.245 0.811) we bring into the operating room

Figure 3.4b **A second example of the Link Grammar Parser Output and the associative TF-IDF weights (INV15B).**

3.3.3 Audio Segmentation and Keyword Extraction

Speech recognition aligns audio with keyphrases selected from the transcript. The keyphrases represent concise words or phrases that correspond to the video content, but their audio counterpart does not always sound clear or legible. The audio may be cropped and fragmented, and will not necessary correspond to natural phrases in human speech. In addition, the quality of speech varies between speakers. Whereas it is easier to segment the voice of a professional speaker, ordinary people do not speak clearly during an interview, and their audio is often mumbled and difficult to segment.

Audio Parsing for Keyphrases

Speech transcription and language characterization were used to detect keyphrases so far. An alternative approach is to detect natural transitions between speakers and topics that are usually marked by silence. A method of locating low energy areas in the acoustic signal detects clear and concise speech regions with more intelligible keyphrases and less emphasis on language characteristics. Audio parsing for keyphrase detection is discussed in chapters 4 and 6.

To detect breaks between utterances, Hauptmann *et al.* [Hauptmann 1995] used the signal power,

$$Power = \log\left(\frac{1}{n}\sum S_i^2\right)$$

where S_i is a low frequency pre-emphasized sample of speech within a frame of 20 milliseconds. A low power level indicates that there is little active speech occurring in this frame. An iterative technique is used to select an appropriate threshold for each video segment. The default power threshold is 25 db and a maximum of 60 iterations at increments of 1 db is allowed. The average duration for a legible audio phrase is 7.5 seconds in documentary video. A threshold is selected when at least 8 audio regions are present for every minute of video in a segment. This corresponds to a desired keyphrase duration of roughly 7.5 seconds (or 8 audio regions per 60 seconds). The average audio duration for phrases differs in different video media, and that fact is used to produce video skims in section 4.3. Speech recognition can then be used to decipher the words in these regions. We rely on automatic speech recognition and closed captions for segmentation of the words in the audio track. Figure 3.5 shows an example of audio utterances detected with noun phrase parsing and audio segmentation.

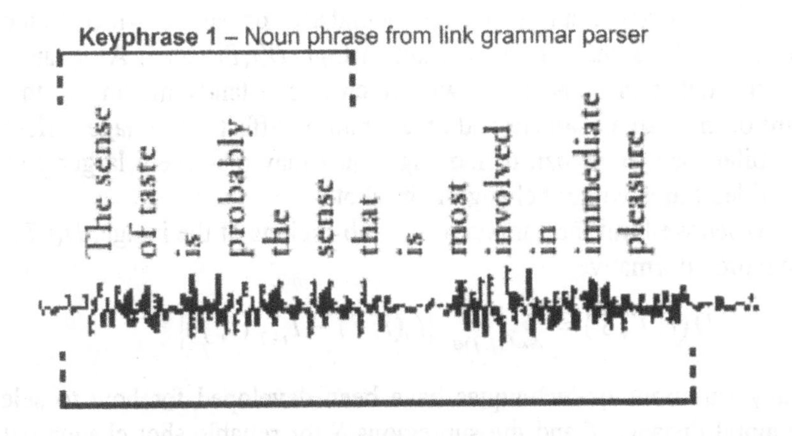

Figure 3.5 Keyphrases - Noun phrase (Keyphrase 1) and audio segmentation phrase (Keyphrase 2). Although the noun phrase detected with the link grammar parser is grammatically correct, it does not convey the accurate meaning of the sentence.

3.4 Image Understanding

As with audio and language understanding for the transcript, there is a collection of useful methods for image understanding of video, including compression techniques for increased storage and faster processing. The sections below describe some of the basic techniques.

3.4.1 Image Difference and Histogram Analysis

Time difference of an image itself or its histogram is the basis for detecting shot changes.

Absolute Image Difference for Shot Change Detection

The simplest difference is the sum of the absolute differences of pixel values. The first image I_t is subtracted by a second image I_{t-T} at a temporal distance T. The difference value is defined as,

$$D(t;T) = \sum_{(i,j)} \left| I_t(i,j) - I_{t-T}(i,j) \right|$$

where the summation is taken over the whole image. When the value of image difference $D(t;T)$ is high, that is interpreted as the shot change.

This method of shot change detection is extremely sensitive to non-essential changes in an image. The boundaries or parts of video frames may contain artifacts that can also result in high $D(t;T)$ value. An example is a running ticker in broadcast news, which is constantly moving in the lower third of the frame. Camera and object motion affects the image difference in a similar manner. A pan or moving object may displace a larger portion of the video frame without changing the shot.

When we limit the summation to sub-regions of the image, $D(t;T)$ values are more informative.

$$D(t;T;S) = \sum_{(i,j)\in S} \left| I_t(i,j) - I_{t-T}(i,j) \right|$$

Many variations of techniques have been developed for how to select the temporal distance T and the subregions S for reliable shot change detection [Arman 1994], [Zhang 1993], [Hampapur 1995], and [Lienhart 1999].

Histogram Difference for Shot Change Detection

A histogram $H(v)$ is a frequency distribution of image value v in an image. It is a simple but useful tool for measuring some properties of images. Figure 3.6 shows examples of grayscale histograms from complex and low interest images. A histogram with concentration to low pixel values or to a particular small range of values indicates an image of "low interest", as it is a low intensity or low contrast image, respectively, which usually conveys little information. Black video frames also signify an abrupt change in subject matter in most documentary and news video. Image frames of other uniform color, however, do not necessarily imply topical changes.

A low-pass filter is often used to eliminate excess noise in the image and subsequent difference, as shown in figure 3.7. The image after filtering with the Gaussian filter is blurred, and its histogram contains fewer colors.

Time difference of grayscale or color histograms can also be computed:

$$D_H(t;T) = \sum_{v=0}^{N-1} \left| H_t(v) - H_{t-T}(v) \right|$$

The histogram difference value, $D_H(t;T)$, will rise due to shot changes, image noise, and camera or object motion. Histogram difference is a simple technique for shot change detection, and yet robust enough to maintain high levels of accuracy for this purpose. The value $D_H(t;T)$, as opposed to image difference $D(t;T)$, is less sensitive to subtle motion, and is an effective measure for detecting shot changes. The temporal spacing T is typically on the order of 5 to 10 frames for video encoded at standard 30 fps. The histogram difference may also be done in sub-regions to limit distortion due to noise and motion.

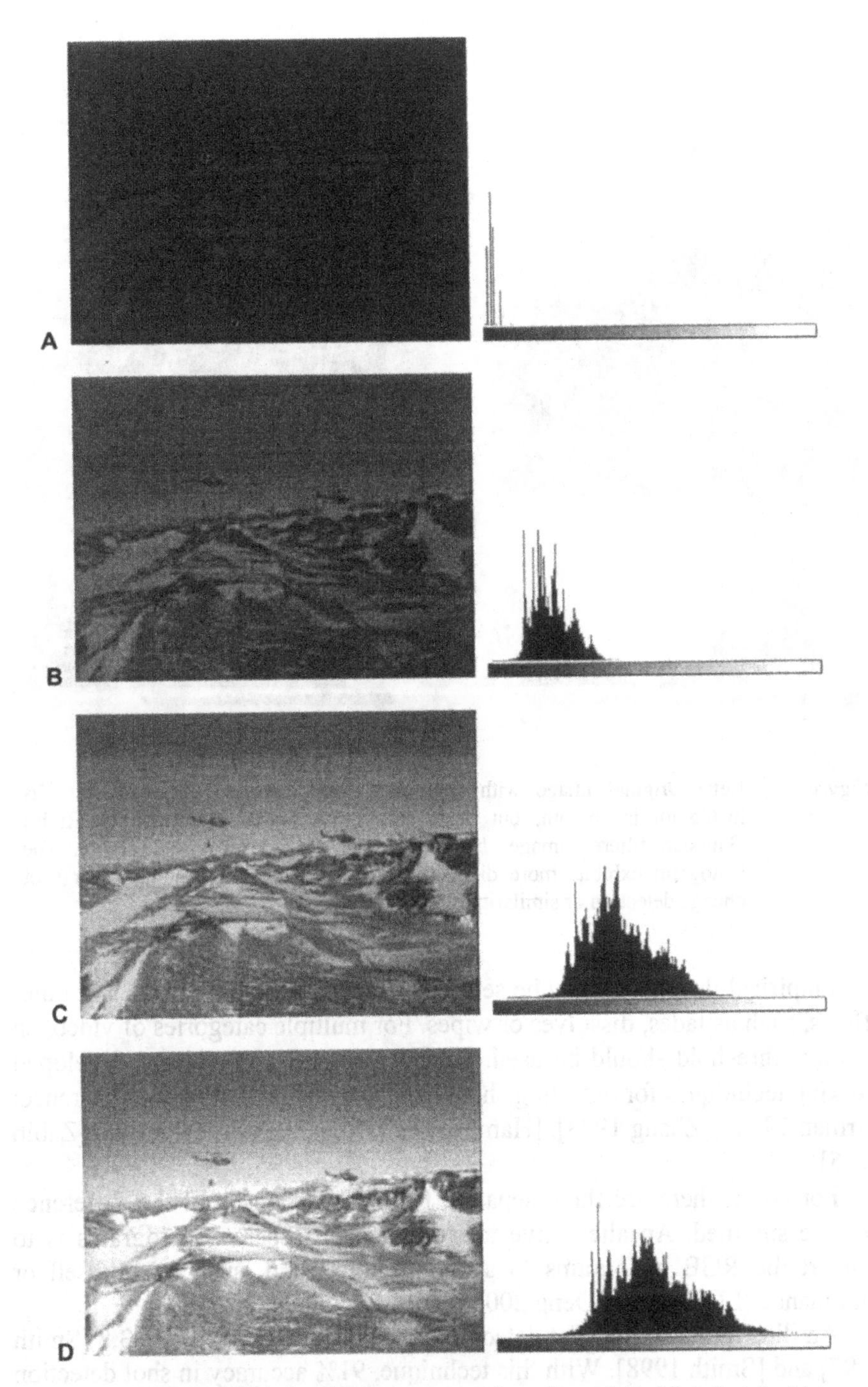

Figure 3.6 Histograms for the dissolve sequence from figure 2.9b. Note that the histogram is normalized such that the peak value is the same. Images C and D have high contrast histograms, and are better for thumbnail representation. Images A and B are dark and contain little visual information.

Figure 3.7 Left) Original image with some noise and encoded edge artifacts. Its histogram is smooth, but the intensity clusters are not distinct. Right) Gaussian filtered image. It is less noisy, but loses visual clarity. The histogram exhibits more distinct clusters, possibly more suitable for shot change detection or similarity matching.

Empirical thresholds may be set that correspond to shot changes or other effects, such as fades, dissolves or wipes. For multiple categories of video, an adaptive threshold should be used. Several research groups have developed working techniques for detecting shot changes by using histogram difference: [Arman 1994], [Zhang 1993], [Hampapur 1995], [Lienhart 1999], and [Zabih 1995].

For color, there are three separate RGB histograms, and the difference may be summed. An alternative to summing the separate histograms is to convert the RGB histograms to a single color band, such as Munsell or Luminance (LUV color) [Deng 2001] and [Manjunath 2001].

An illustration of the shot detection results is shown in Figure 3.8 [Smith 1997] and [Smith 1998]. With this technique, 91% accuracy in shot detection was achieved on a test set of roughly 495,000 images (5 hours). MPEG-1 video is segmented at 36 fps on an SGI Indigo 2 workstation (MIPS R4400 200 MHz). Table 3.2 lists the performance statistics of shot change detection.

Table 3.2 Shot Change Detection Results - L1 norm histogram difference measure with quad sub-regions.

Video Segment (Duration-30 fps)	Shots Detected	Shots Missed	False Shots
Infinite Voyage Intro. (535 seconds)	38	1	0
Underwater Exploration (738 seconds)	34	3	4
CNN News I (1422 seconds)	131	16	5

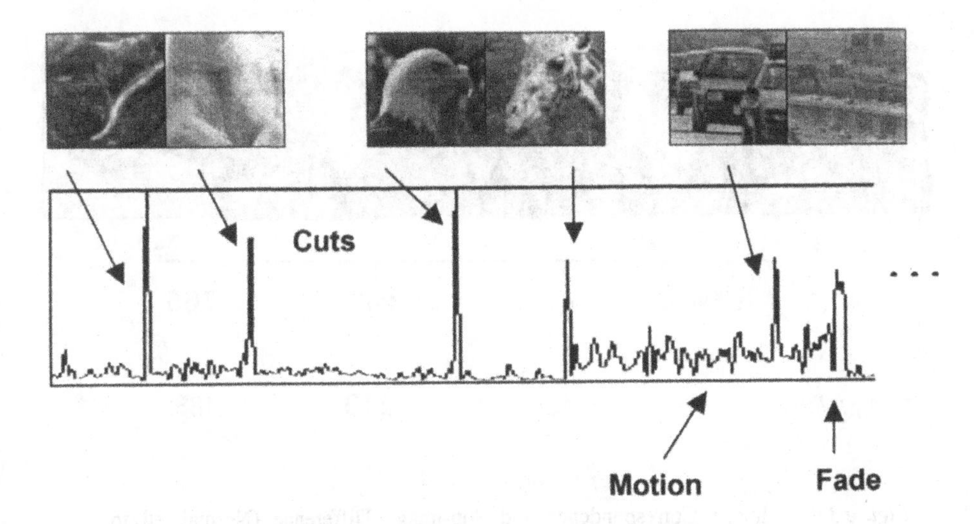

Figure 3.8 Histogram-based image difference $D_H(t;T)$, for shot detection. Fades, dissolves, wipes and other visual effects usually appear as broader peaks. Camera and object motion, motion blur and noise often cause shorter successive peaks. Cuts are the predominant shot change and usually appear as sharp peaks.

3.4.2 Histograms for Segmentation and Image Correspondence

Image correspondence is important for identifying frames or scenes that appear throughout a segment. In news footage, the anchorperson usually appears with a similar pose and in the same background for different stories in a video. The appearance of the anchorperson, if detected, can then be used to delineate story breaks for temporal segmentation. In documentaries, a person being interviewed will appear at various points throughout the story and especially when topical changes take place.

Color histograms can be used as a measure for image correspondence. A histogram of a video frame is compared with that of video frames in subsequent shots. Without prior analysis, the first, middle or some other fixed frame are selected from the shot for comparison. This method can select dark or unintelligible images. Histograms are used to verify visible contrast levels as shown in Figure 3.6. The histogram difference described here is widely used: [Gong 1994], [Zhang 1995], and [Yeung 1995].

An example of an image correspondence result is shown in Figure 3.9. As the scene progresses, $D_H(t;T)$ remains low, even when a graphical logo is placed in the shot.

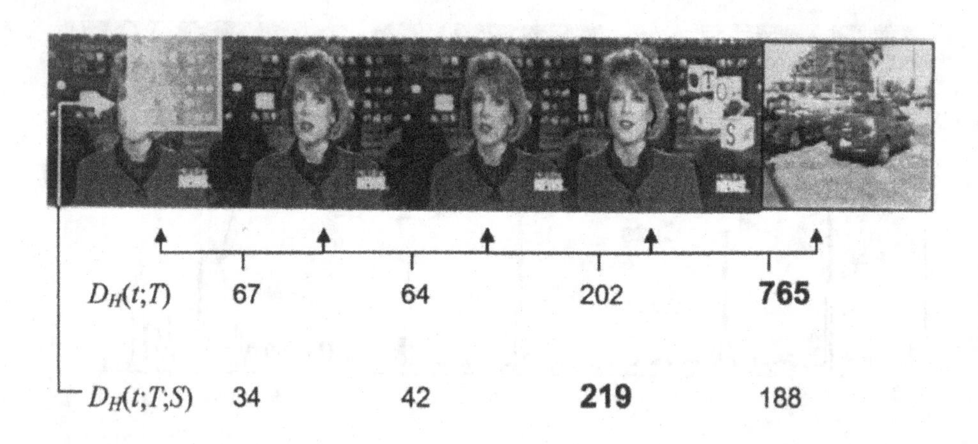

Figure 3.9 Image Correspondence and Sub-image Difference (Normalized to 1000): $D_H(t;T)$, Global image histogram difference with a shot change above 500. $D_H(t;T;S)$, Sub-image histogram difference with graphics detected above 100 and $D_H(t;T)$ below 500. Images 1 through 3 are replicated for this illustration.

Sub-Image Histogram Difference for Topic Change Detection

In news footage, an icon or logo is often used to symbolize the subject of the video. An anchorperson sometimes narrates multiple segments while corresponding news icons appear in the background. This icon is usually placed in an upper-quarter of the image. Although the background of the image remains the same, changes in this icon represent changes in content. By applying the histogram difference to a small region S in the image, we can detect changes in news icons. A fixed detection window is applicable when the video format does not change.

In Figure 3.9, $D_H(t;T;S)$ is the sub-region histogram difference of the upper-right quadrant of the frames. $D_H(t;T;S)$ increases during shot changes and the insertion of a graphic logo. If $D_H(t;T;S)$ is above the empirical threshold, and no shot change is detected, a sub-region change is marked. Figure 3.10 shows a close-up of the icon change detection in Figure 3.9. A moving sub-image window is needed when graphics are placed at multiple locations on the screen.

Frame *t* **Frame *t+T***

Figure 3.10 Topic Change Detection by Sub-Region Histogram Difference. In this example, the upper-right quadrant is known to change during the introduction of a news story. The sub-image window is offset from the image borders by 10 pixels to avoid edge artifacts.

3.4.3 Texture and Edge Features

Texture distinguishes images of low complexity from those containing busy image features. A texture measure is computed from perceptual image features, such as coarseness, contrast, directionality, and regularity [Bovik 2000], [Shapiro 1992], and [Gonzalez 1992]. An image with limited texture is often low contrast and uniform in color. Texture can also be used as a measure for thumbnail selection. Image blur, often introduced as a visual effect, yields a low textured display.

Texture is useful for image segmentation where color segmentation methods fail. Coarse images, like the example in Figure 3.11a, are difficult to segment with only pixel-wise color analysis. The segmentation result to the right was obtained by coupling the boundaries of an edge map and the color regions. Segmented images like this are used in content-based image retrieval (CBIR) to match individual objects or regions; texture features are combined with color for improved image retrieval [Cohen 1999], [Iqbal 2002], [Manjunath 1996], [Rubner 1998], and [Rubner1999].

Texture is also used as a measure of coarseness, as seen in Figure 3.11b. A homogeneous image will exhibit much less texture than a coarse image. In this figure, an edge-based measure was used to create the edge maps below the images. Texture percentage is based on the number of edge pixels in the binary edge-map.

Similar edge features were used by Zabih to detect shot changes [Zabih 1995]. Edge features are detected and tracked over time. The edge map remains constant in static video and moves with camera or object motion. At shot changes, these features fade or disappear, except for shot changes by morph effects.

Figure 3.11a Left Image) Texture representation generated from an edge map; Right Image) Segmented regions.

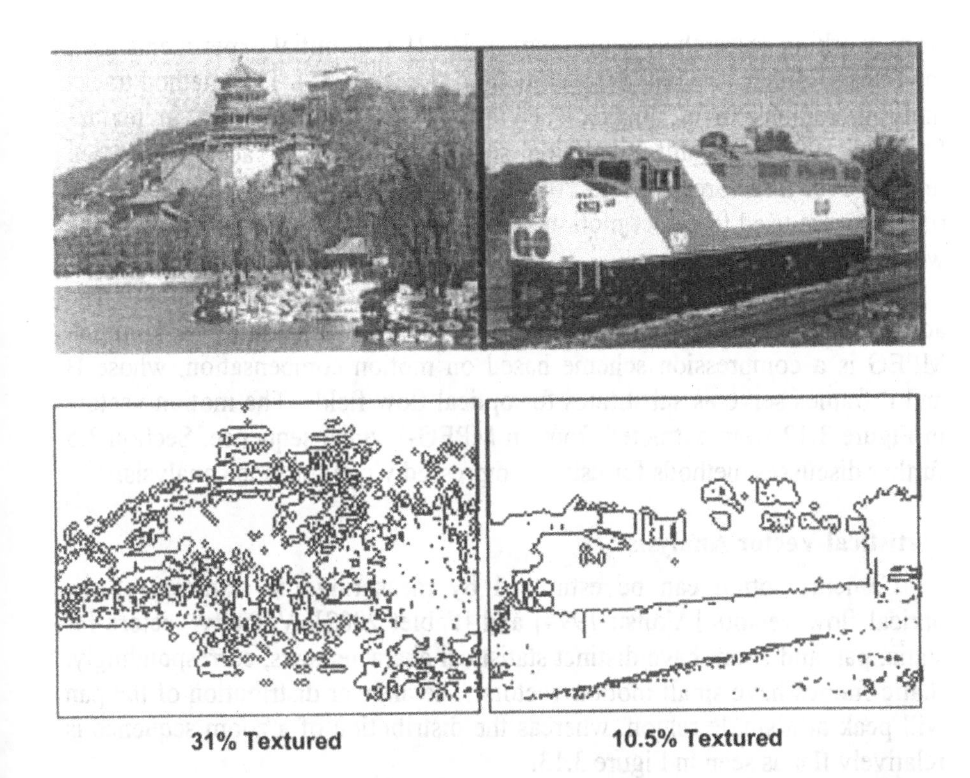

Figure 3.11b Left Image) Coarse image with texture representation generated from edge map - 31% textured (black portion of edge map); Right Image) Smoother image with less texture in the corresponding edge map - 10.5% textured.

3.4.4 Camera Motion

An important aspect of video characterization is camera motion [Tse 1991] and [Fablet 2002]. Many scenes have beautiful motion effects, but offer little in the description of a particular segment. Static scenes, such as interviews and still poses, contain several redundant video frames. For video summarization applications, we wish to eliminate video with excess camera motion or visual redundancy. We first describe optical flow detection methods and statistical camera motion analysis. We then describe faster methods for optical flow using compressed video, and affine models for classifying camera motion.

Motion Vectors

Motion vectors provide the basis for camera motion detection. Most methods for generating motion vectors track individual regions from one video frame to the next to create optical flow fields like the example in Figure 3.12. Barron's survey of optical flow analysis provides an overview of many working research systems [Barron 1992]. Our initial experiments used the Lucas-Kanade gradient descent method [Lucas 1981]. This method tracks multiple regions in an image, such as corners, or areas rich in texture [Tomasi 1990]. A multi-resolution structure is used to accurately track regions over large areas and to efficiently compute optical flow. The top 30 regions were used to detect motion rather than computing optical flow in the whole image.

Compressed data of video can be used for obtaining optical flow. The accuracy of the optical flow is reduced, but the computation is minimal. MPEG is a compression scheme based on motion compensation, whose B and P frames serve as substitutes for optical flow fields. The motion vectors in Figure 3.12 were extracted from an MPEG-1 video sequence. Section 3.5 further discusses methods for using compressed video for video analysis.

Statistical Vector Analysis

Camera motion can be estimated by measuring the statistics of the optical flow vectors [Akutsu 1994] and [Fablet 2002]. Velocity vectors for static, pan and zoom have distinct statistical characteristics, correspondingly. Static frames have small motion vectors. The angular distribution of the pan will peak at a single region, whereas the distribution of a zoom sequence is relatively flat as seen in Figure 3.13.

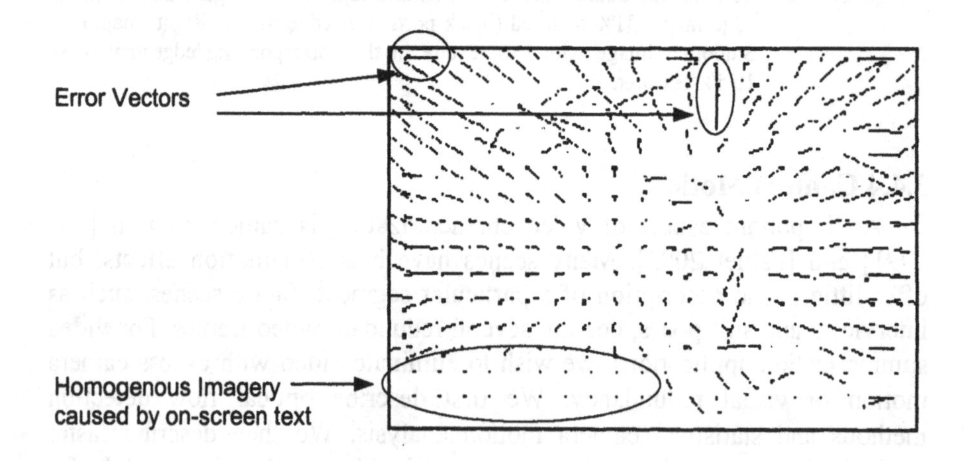

Figure 3.12 Motion vectors extracted from MPEG-1 video sequence. Error vectors and vectors near homogeneous regions contribute to the motion vector confidence. This sequence is a close-up of the Zoom shot in Figure 3.14.

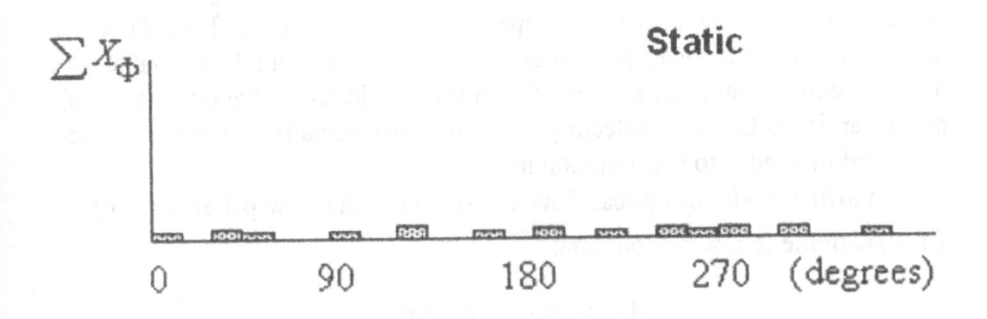

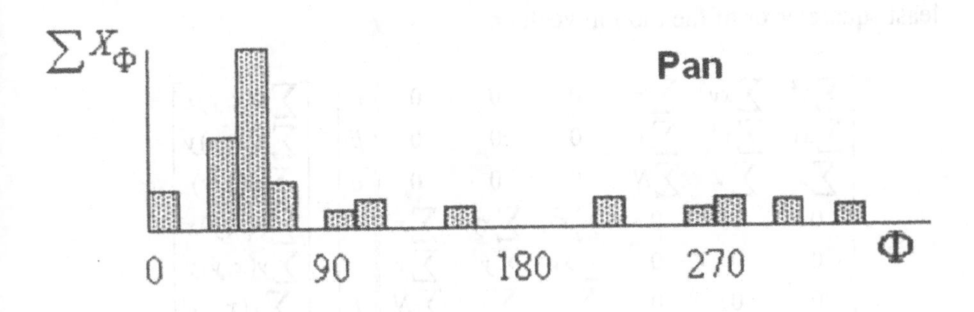

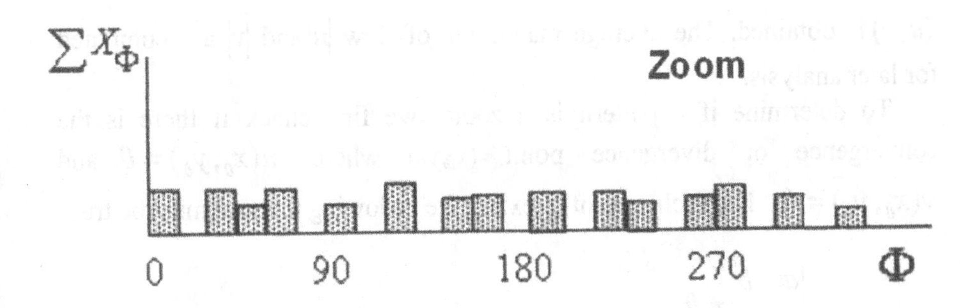

Figure 3.13 Angular distributions for different forms of camera motion analysis.

Affine Motion Model

The last section described an intuitive method to distinguish camera motion based on the statistical properties from the optical flow. Though useful, use of statistical properties alone has been found unreliable in distinguishing a pan from a zoom. The ability to identify the zoom shot is of particular importance in selecting the most representative video. A more advanced method is to use a motion model.

An affine model of optical flow approximates the flow patterns (u,v) at (x,y) with the following equation:

$$u(x,y) = ax + by + c$$
$$v(x,y) = dx + ey + f$$

Affine parameters *a b, c, d, e,* and *f* are calculated by minimizing the least square error of the motion vectors.

$$\begin{bmatrix} \sum x^2 & \sum xy & \sum x & 0 & 0 & 0 \\ \sum xy & \sum y^2 & \sum y & 0 & 0 & 0 \\ \sum x & \sum y & \sum N & 0 & 0 & 0 \\ 0 & 0 & 0 & \sum x^2 & \sum xy & \sum x \\ 0 & 0 & 0 & \sum xy & \sum y^2 & \sum y \\ 0 & 0 & 0 & \sum x & \sum y & \sum N \end{bmatrix} \begin{bmatrix} a \\ b \\ c \\ d \\ e \\ f \end{bmatrix} = \begin{bmatrix} \sum u(x,y)x \\ \sum u(x,y)y \\ \sum u(x,y) \\ \sum v(x,y)x \\ \sum v(x,y)y \\ \sum v(x,y) \end{bmatrix}$$

The summation is taken over all the positions (x,y) at which the flow vector (u,v) is obtained. The average magnitude of flow $\overline{|u|}$ and $\overline{|v|}$ are computed for later analysis.

To determine if a pattern is a zoom, we first check if there is the convergence or divergence point (x_0, y_0) where: $u(x_0, y_0) = 0$ and $v(x_0, y_0) = 0$. For such a point to exist, the following relation must be true:

$$\begin{vmatrix} a & b \\ d & e \end{vmatrix} \neq 0$$

If the above relation is true, then compute (x_0, y_0), the focus of expansion. If the computed (x_0, y_0) is located inside the image, and $\overline{|u|}$ and $\overline{|v|}$ are large, then the camera is zooming. If (x_0,y_0) is outside the image, and $|\overline{u}|$ and $|\overline{v}|$ are large, then the camera is panning in the direction of the dominant vector.

If the above determinant is approximately 0, then a convergence or divergence point (x_0,y_0) does not exist and the camera is panning or static. If $|\overline{u}|$ or $|\overline{v}|$ are large, the motion is panning in the direction of the dominant vector. Otherwise, there is no significant motion and the flow is static.

Averaging the results in a 20-frame window over time eliminates erroneous detection due to fragmented motion. We used the following percentages of frames per motion category needed for classification as such.

- Pan Motion - 70%
- Zoom Motion - 60%
- Static Motion - 75%
- Random Motion - 50%

The processing rate on an SGI Indigo 2 (MIPS R4400 200 MHz) was 26 fps. Examples of the camera motion analysis results are shown in Figure 3.14. Table 3.3 shows the statistics for detection on various image sets (regions detected are either pans or zooms).

Table 3.3 Camera Motion Detection Results

Data (Frames Processed)	**Regions Detected**	**Regions Missed**	**False Regions**
Destruction of Species I - II (20724 Frames)	23	5	1
Planet Earth I-II (25680 Frames)	36	1	3
CNHAR News (30520 Frames)	14	1	2

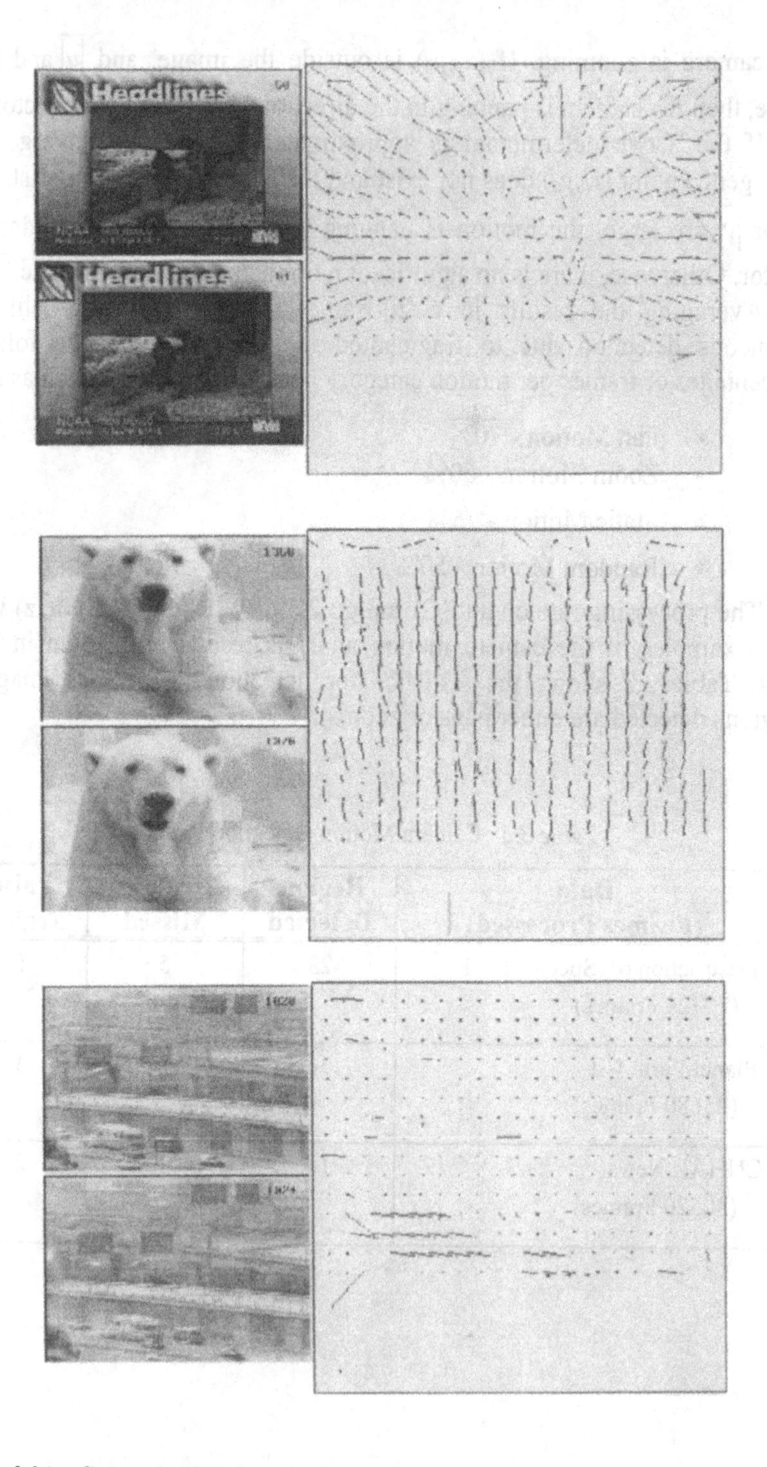

Figure 3.14 Camera motion from MPEG-1 motion vectors: Top) Zoom sequence, Middle) Pan sequence, and Bottom) Static sequence with minor object motion from a moving bus in the lower left corner.

3.4.5 Motion for Segmentation

Camera motion provides a tool for video segmentation and shot change detection. Fast camera motion is often falsely misrecognized as a shot break when only the histogram distribution is measured. The camera motion estimate can be used to set shot change thresholds adaptively for the histogram difference measure, $D_H(t;T)$. Figure 3.15 shows a comparison of the histogram-based shot change detection and shot changes detected from a motion vector confidence measure. The motion-based method is more sensitive to movement, but the dominant peaks correlate well with the histogram method. In this example, confidence is based on the tracking metric from Lucas [Lucas 1981].

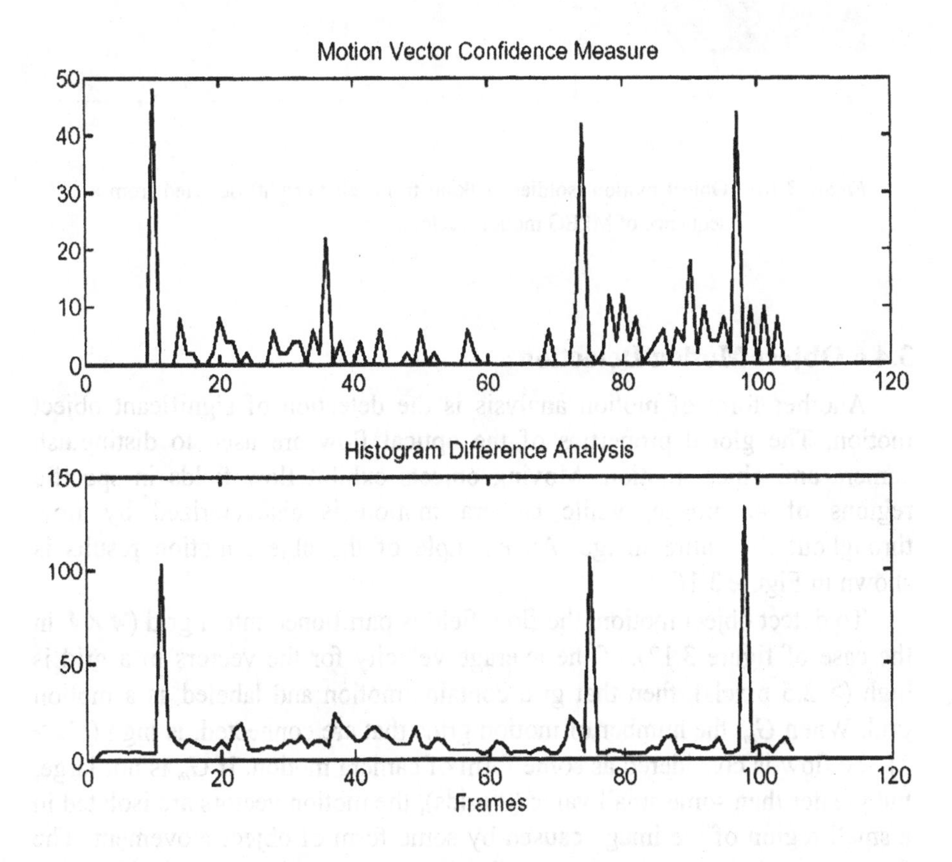

Figure 3.15 **Top) Shot detection from a motion vector confidence measure, and Bottom) Histogram-based Shot Detection. The sequence between frames 15 and 70 is a tracking shot of a moving animal.**

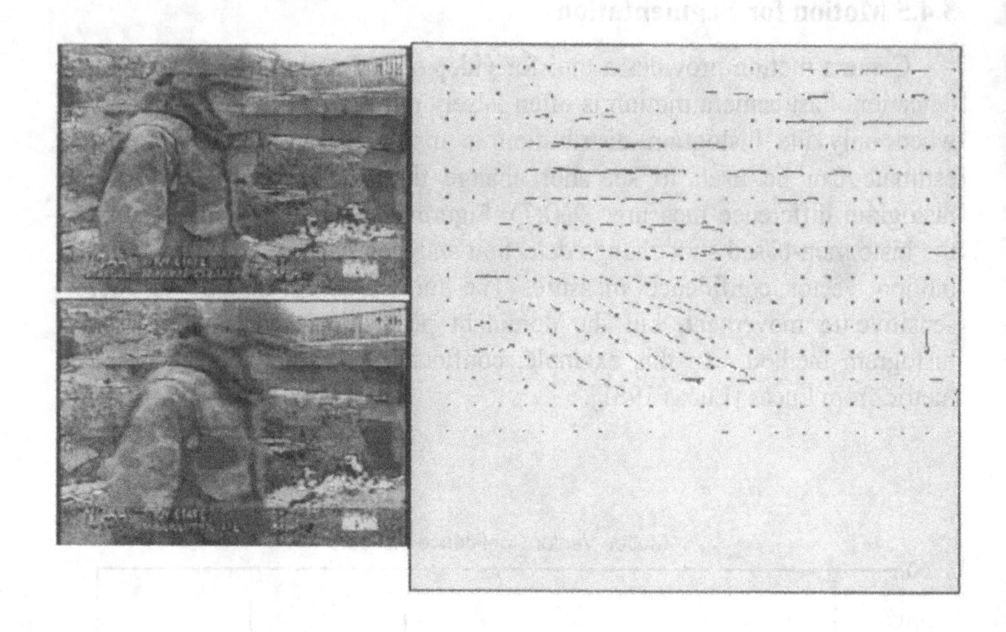

Figure 3.16 Object motion (soldier walking from left to right) detected from a sequence of MPEG motion vectors.

3.4.6 Object Motion Detection

Another form of motion analysis is the detection of significant object motion. The global properties of the optical flow are used to distinguish camera and object motion. Moving objects exhibit flow fields in specific regions of an image, while camera motion is characterized by flow throughout the entire image. An example of the object motion results is shown in Figure 3.16.

To detect object motion, the flow field is partitioned into a grid (4×4 in the case of figure 3.17). If the average velocity for the vectors in a grid is high (> 2.5 pixels), then that grid contains motion and labeled as a motion grid. When G_m, the number of motion grids that are connected, is high (G_m > 7), the flow is considered as some form of camera motion. If G_m is not large, but greater than some small value (2 grids), the motion vectors are isolated in a small region of the image caused by some form of object movement. The frame is labeled as motion frame. Like the camera motion analysis, if motion frames are persistent over a 20-frame window (more than 60%), the sequence is considered as object motion. An illustration of this algorithm is shown in Figure 3.17.

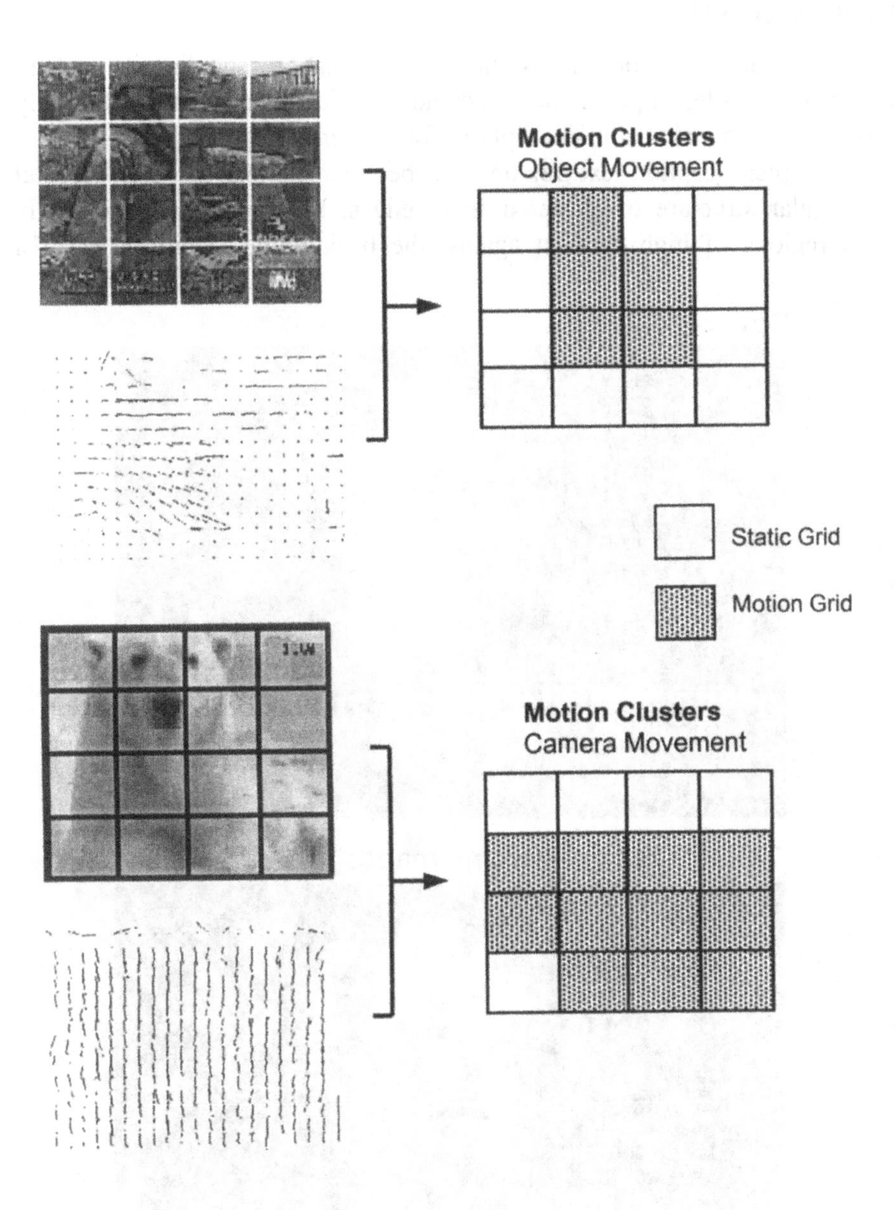

Figure 3.17 Illustration of object motion detection. Top) Image from Figure 3.16 containing a walking soldier. Motion clusters are centered in a localized region. Bottom) Image from a pan sequence with global motion clusters.

3.4.7 Caption Detection

Text in the video provides significant information regarding its content. For example, titles of people involved and statistical numbers are not usually spoken, but are included in the captions for viewer inspection.

A typical caption-text region can be characterized as a horizontal rectangular structure of clustered sharp edges, because characters usually form regions of high contrast against the background [Smith 1997], [Li

Figure 3.18 Image of time-based minimum tends to sharpen the area of video caption, while blurring the rest.

2000], and [Lienhart 2001]. By detecting these properties, we can extract regions from video frames that may contain textual information. Consistent detection of the same text region over a period of time is probable since text regions remain at an exact position for many video frames. The on-screen text does not move and maintains a near constant level of clarity. Contrast enhancement of characters in a text region can be achieved by time-based minimum operation – taking the minimum or darkest pixel value over a period of time. Figure 3.18 shows an example.

Figure 3.19 illustrates the process of text detection - primarily, regions of horizontal titles and captions. We first apply a horizontal differential filter to the entire image, and with appropriate binary threshold extract vertical edge features.

$$\textit{Differential Filter} \quad \begin{bmatrix} 0 & 1 & 0 \\ -\frac{1}{2} & 1 & 1/2 \\ 0 & 1 & 0 \end{bmatrix}$$

Smoothing filters are then used to eliminate extraneous fragments, and to connect text pixels that may have been detached. The size of this filter is relative to the font size. Individual regions are identified by cluster detection, and their bounding boxes are computed.

We now select clusters with bounding boxes that satisfy certain properties of cluster size, cluster fill-factor, and horizontal-vertical aspect ratio of the bounding box. Cluster size C_S is the number of pixels in the cluster, and cluster fill-factor C_{FF} is the ratio of C_S to the area of the bounding box B_A:

$$C_{FF} = \frac{C_S}{B_A}$$

For a cluster to be a text region, its bounding box must have a small vertical-to-horizontal aspect ratio as well as satisfying various limits in height and width. The fill factor should be high to insure dense clusters. The cluster size should also be relatively large to exclude small fragments from noise. These controlling parameters are listed below.

- Maximum Cluster Height = 50 pixels
- Minimum Cluster Height = 10 pixels
- Maximum Cluster Width = 150 pixels
- Minimum Cluster Width = 15 pixels

Cluster Size > 70 pixels

Cluster Fill Factor density ≥ 0.45

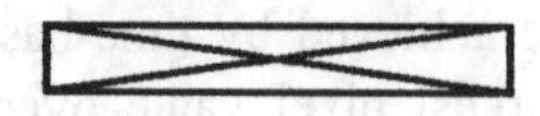

Horizontal-Vertical Aspect Ratio ≥ 0.75

Finally, the intensity histogram of each region may be examined to test for high contrast. This is to eliminate certain textures and shapes that appear similar to text but exhibit low contrast. This histogram test is not needed with most grayscale text. Color text often exhibits less contrast and often requires a histogram test for accurate detection.

Figure 3.20a and 3.20b show several detection examples of words and subsets of words. The performance is to detect over 94% of more than 140 text regions contained in 250 images, while producing roughly 34 false detections. Table 3.4 lists the text detection statistics for several experiments.

To read the text in the detected text region, a standard optical character recognition (OCR) package may be used. For most OCR systems, the input is assumed to be an individual character. This requires segmentation of the text region into individual character regions; the segmentation is not an easy problem in digital video since most of the characters experience some degradation during recording, digitization and compression. For a simple font, we can search for blank spaces between characters and assume a fixed width for each letter. For the case of unknown and multiple fonts, certain heuristics are useful to obtain better results than generic OCR systems [Sato 1998]. Figure 3.21 shows an example of these results for the Time New Roman and Helvetica fonts. With the exception of the word "County" due to the scene texture overlap, the phrase is recognized accurately.

Table 3.4 Text Detection Results [Smith 1997]

Data (Frames)	Regions Detected	Regions Missed	False Regions
CNHAV News (1056 frames)	26	1	3
CNHAR News (1526 frames)	48	0	5
Destruction of Species I (264 frames)	12	2	0
Planet Earth I-II (1712 frames)	0	0	2

Input Image

A single frame is extracted from a video or image sequence. When successive images are averaged over time at intervals of 2 to 3 frames, the background region becomes blurred. In most cases the text remains clear, while the background is blurred.

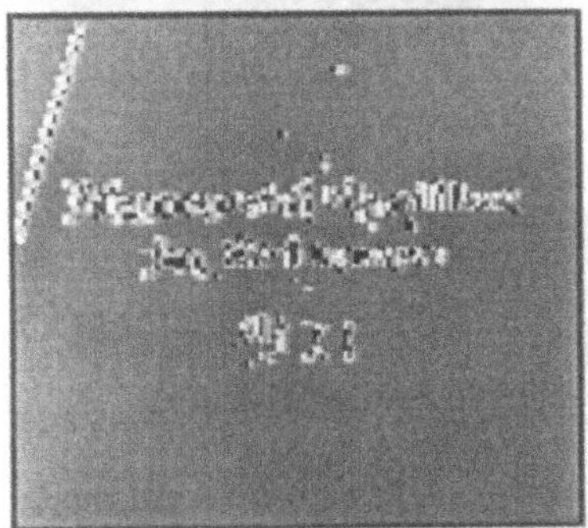

Differential Filter

The differential filter extracts the majority of the components that resemble text. Undesirable artifacts such as straight vertical lines, noise and random occurrences of high frequency also appear after filtering.

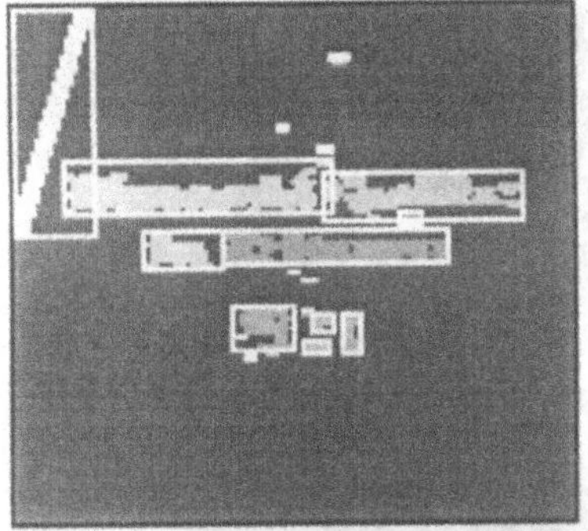

Clustering and LPF

A horizontal low-pass filter is applied for smoothing. These regions are then clustered into individual lumps and associated bounding boxes. The differential filter eliminates most sparse artifacts and many algorithms may be effectively used for clustering [Jain 1989] and [Shapiro 1992].

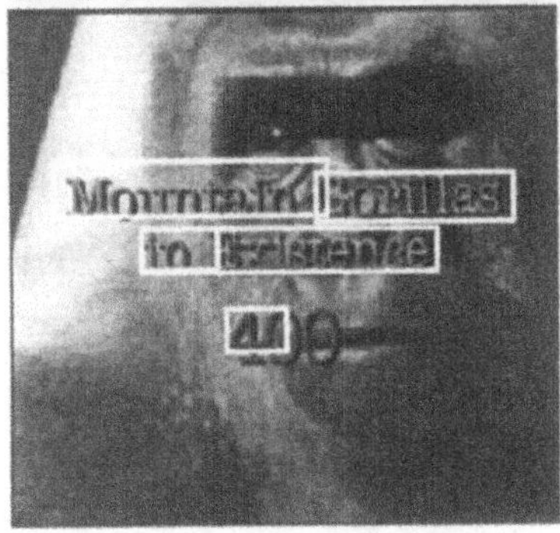

Region Analysis

Clustered regions that do not satisfy the criteria mentioned above are eliminated. The final output displays the text regions and may be used for purposes such as indexing and OCR.

Figure 3.19 Text region extraction: filtering, clustering, and region analysis.

Figure 3.20a Examples of text and face detection

Figure 3.20b Examples of text and face detection

Figure 3.21 OCR from detected text regions. Fixed fonts, such as Times New Roman (SHERMAN BLOCK – Top Line) and Helvetica (LOS ANGELES COUNTY SHERIFF – Bottom Line) are used in broadcast news and other venues. OCR systems can be trained to recognize a variety of fonts.

3.4.8 Object Detection

Identifying significant objects in video is one of the key capabilities in video characterization. A number of programs have been developed for the detection of a particular object, such as human faces, text, or automobiles [Shapiro 1992], [Furht 1996], [Furht 1999], and [Schneiderman 2002]. These systems targeted for specific objects have higher accuracy than systems that attempt to identify all objects in the image. It has also been demonstrated that perceptual grouping results in recognition of a small group of objects, such as certain four-legged mammals, flowers, specific terrain, clothing, and buildings [Forsyth 2000], [Iqbal 2001], and [Iqbal 2002].

Human face detection is a common theme in video characterization. The talking-head image is widely used in video productions of interviews and news clips, and is a clear indication of focus on an individual of interest. Most detection methods use pattern matching to locate human faces [Sung 1998]. Facial templates based on distinct local regions such as the eye, nose and mouth are applied throughout an image. Recent work has addressed problems with variations of scale, rotation, orientation and minor occlusion. One example is a system developed at Carnegie Mellon University by Rowley, Kanade and Baluja [Rowley 1998]. Their neural network arbitration method for human-face detection was invariant to scale and rotation, and was especially reliable with frontal faces such as talking-head images. Figure 3.22 shows examples of its output, illustrating the range of facial size and rotation. The newer program by Schneiderman and Kanade [Sneiderman 2002], based on a Baysian classification method, detects faces up to at 90-degree out-of-plane rotation and shows further improvement, as well as generalizing applicability to other objects, such as cars. Figure 3.23 shows examples of its results. The performance statistics of these programs is listed in Table 3.5.

Future challenges in face detection include poor lighting conditions, partial occlusion and facial orientation. Recent interest in biometric and security applications has contributed to improved methods in face detection and recognition [Biometrics 2003].

Table 3.5 Face Detection Results on FERET images [Schneiderman 1998].

	Schneiderman and Kanade		Rowley, Baluja, and Kanade	
Data set	Detection Rate	False Alarms	Detection Rate	False Alarms
0 ° set	99.6%	1	98.7%	3
15 ° set	100.0%	0	99.6%	0
22.5 ° set	99.7%	2	95.5%	3

Figure 3:22 Examples of face detection at different sizes and facial rotation [Rowley 1998].

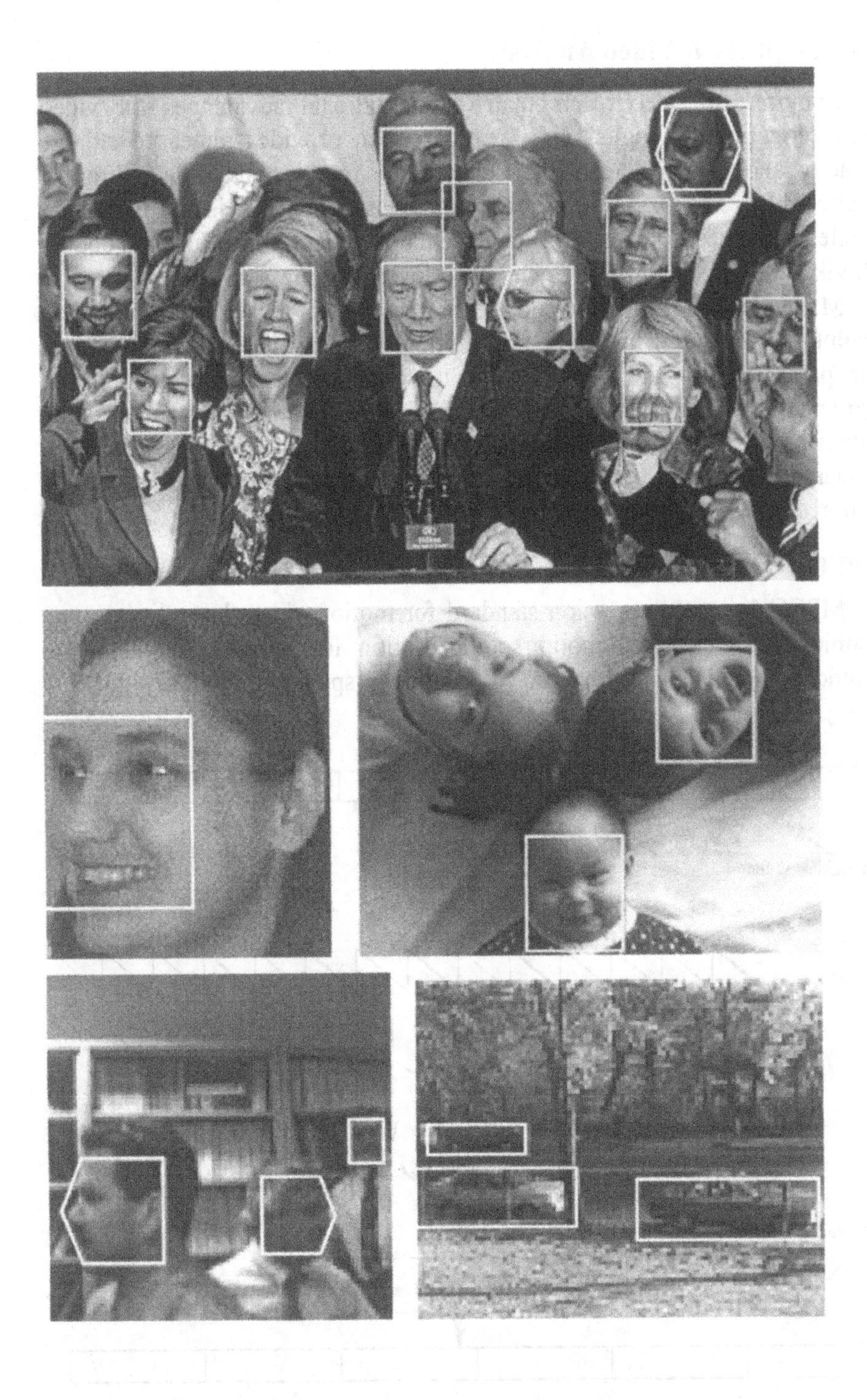

Figure 3:23 Detection of rotated faces and automobiles [Schneiderman 2002]

3.5 Compressed Video Analysis

Compressed video offers many computational advantages for video characterization. In digital video, compression provides increased storage capacity and statistical characteristics of the image track as a byproduct [Meng 1996]. The major drawback to compression schemes is loss in quality. Lossless schemes, such as Run Length Encoding (RLE) and Huffman coding [Bovik 2000] and [Jain 1989], achieves a lower compression rate.

Many algorithms provide compression as high as 100 to 1, depending on resolution, and often use DCT and motion compensation for compression. The parameters of the DCT may be used for shot change detection and segmentation [Zhang 1995] and [Arman 1994]. The motion compensation statistics may be used as a form of optical flow, as discussed in section 3.4.4. Researchers have also developed efficient methods for face detection in compressed video [Wang 1997].

MPEG-1 Video Standard

MPEG-1 is a compression standard for motion video that relies on two techniques: block-based motion compensation for reducing the temporal redundancy and transformation for reducing spatial redundancy [MPEG 2003].

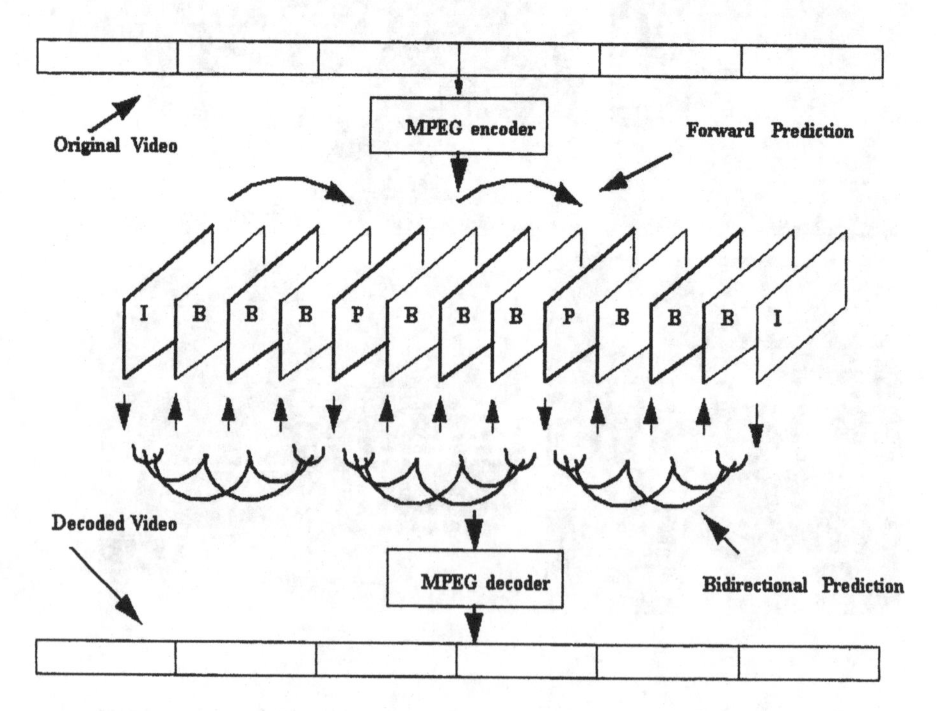

Figure 3.24 Graphical description of the MPEG-1 encoding process.

The MPEG compression algorithm assumes the locality; that is, a current image may be modeled as a displacement of the image at some previous time. The amplitude and direction of the displacement need not be the same everywhere in the image. MPEG does not compress every single frame in the encoding process. It compresses some frames and the locality property of images allows MPEG to encode just the part of an image exhibiting a change in motion.

There are three major types of frames in MPEG coding: intra-picture (I), predicted-picture (P), and interpolated-picture (B). These frames are illustrated in Figure 3.24. To achieve a reduction in spatial redundancy, I frames are compressed by dividing an image into a set of 8 x 8-pixel blocks. The intensity of each block is then transformed by the Discrete Cosine Transform (DCT) into 64 coefficients, which are then quantized and run-length encoded with Huffman entropy. To reduce the temporal redundancy, the motion compensation encoding is applied to P and B frames. P frames are encoded with reference to the past (I or P frames). P frames are then used as a reference for future frames. B frames are encoded with reference to the past and the future. An Intra (I) frame is coded by a block-based (DCT) algorithm with no motion compensation. Motion compensation in a bidirectional (B) frame is interpolated between the forward and backward prediction, so it may accumulate some error. The motion vectors from predicted (P) frames are coded using motion compensated prediction (forward motion vector) from the past reference image.

In the decoding process, for each P frame, the value of forward motion vector for each macroblock is reconstructed. Each forward motion vector consists of two components: horizontal and vertical. If a value of the reconstructed horizontal motion vector is positive, then the referenced area of the past reference image is to the right of the macroblock in the coded P frame. If it is zero, then the referenced area of the past reference image is at the same position of the macroblock in the coded P frame. Likewise, the reconstructed vertical motion vector indicates the motion of the referenced area of the past reference image in the coded image [Tse 1991]. These forward motion vectors can be used as the optical flow vectors, as seen in section 3.4.4. As a result, unlike some algorithms that track only obvious features such as corners and textured regions [Tomasi 1990], MPEG motion vectors appear as a grid since they use each macroblock in a frame as a feature.

Shot Change Detection

The detection of shot changes in compressed video has been explored by many researchers [Nam 2000], [Boccignone 2000], [Gargi 2000], and [Yeo 1995]. These systems typically have demonstrated success at near 95% accuracy (similar to histogram and edge-based methods). They rely on compressed domain features such as DCT components and frame size. Shot change detection in a compressed stream requires less computing time than methods that decode the image stream and then process the decoded images. The histogram difference and edge-based methods require that a compressed stream be decoded for every frame that is analyzed. The DCT coefficients of a MPEG or Motion-JPEG sequence can be used to detect shot changes by observing only the relative energy over time. In most cases, however, the DCT method does not achieve the same accuracy as histogram difference methods.

Another approach to shot change detection is to monitor fluctuations in the size of the compressed frames. Shot changes cause drastic changes in the structure of the MPEG GOP (Group of Pictures). When they coincide with a B or P frame, the size of the B or P frame increases to roughly the size of an I frame. Figure 3.25 shows an example of the size distribution of different MPEG frames.

Video Captions

Text detection is another form of image analysis that can utilize a compressed stream. A modified DCT energy function acts as a filtered component similar to that described in section 3.4.7 [Zhong 2000]. Similar clustering and region-based restrictions may be used to further identify text. A final step involves dilation of the results to extract the full text region. Figure 3.26 illustrates an example of this technique for an image containing multiple text regions.

The MPEG-7 Standard

This chapter has focused on automatic and manual methods for extracting features relevant to video characterization from original or compressed video. It is desirable for video characterization that such features or metadata are available as a standard [Smith and Chen 2000]. MPEG-7 is an ongoing effort by the Moving Picture Experts Group that is working toward this goal, i.e., the standardization of metadata for multimedia content indexing and retrieval [MPEG 2003].

Picture Type	Bit Allocation 30 Hz SIF @ 1.15 Mbit/sec	Bit Allocation 30 Hz ITU-R 601@ 4 MBit/sec
Intra	150 Kbit	400 KBit
Predictive	50 Kbit	200 KBit
Bi-directional	20 Kbit	80 KBit

Figure 3.25 MPEG's I, B, P frame size.

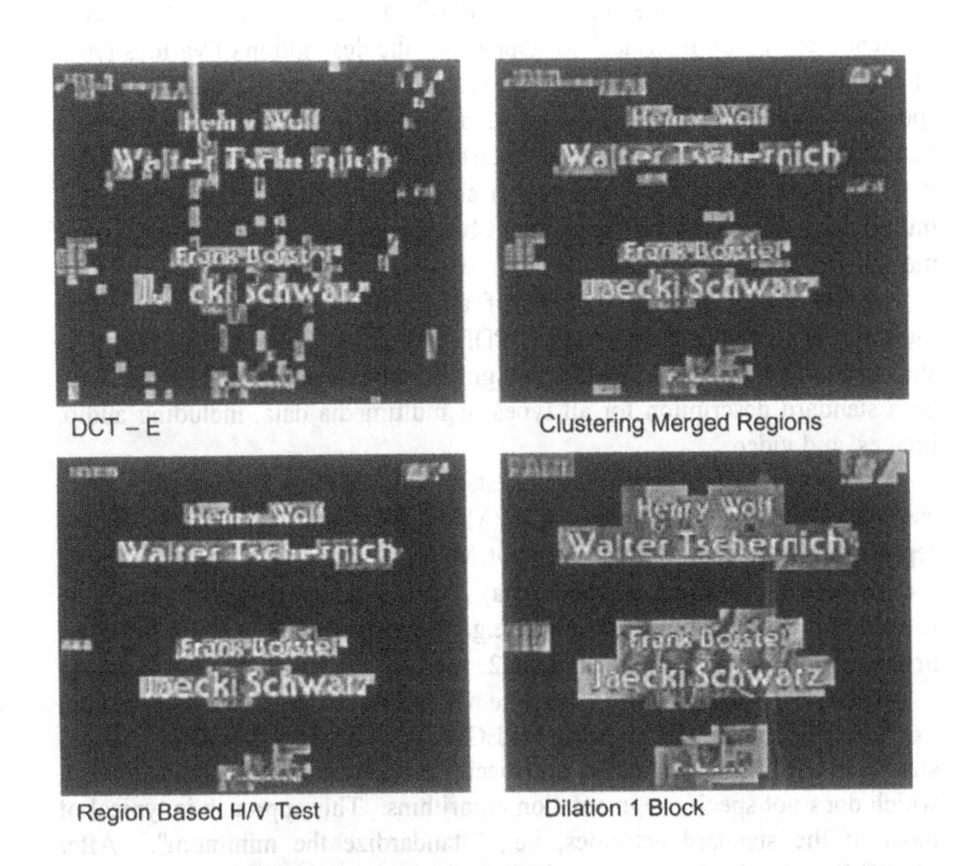

Figure 3.26 Text detection from MPEG DCT-E coefficients, clustering, region selection and pixel dilation [Zhong 2000]. The original image is shown on the bottom.

The goal of MPEG-7 is to enable efficient characterization and descriptions of multimedia content. MPEG-7 strives to define a "Multimedia Content Description Interface" to standardize the description of various types of multimedia content, including still pictures, graphics, 3D models, audio, speech, video, and composition information. It may also support special cases such as facial expressions, and personal characteristics. Once finalized, it will transform the text-based search and retrieval (e.g., keywords) of the multimedia databases into a content-based approach, e.g., using color, motion, or shape information.

MPEG-7 can also be thought of as a standardization of describing multimedia content. If one regards PDF (Portable Document Format) as a standard language to describe text and graphic documents, then MPEG-7 will be a standard description for all types of multimedia data, including audio, images, and video.

Compared with earlier MPEG standards, MPEG-7 possesses some essential differences. For example, MEPG-1, 2 and 4 all focus on the representation of audiovisual data, but MPEG-7 will focus on representing the "metadata" (information about data). MPEG-7, however, may utilize the results of previous MPEG standards, e.g., the shape information in MPEG-4 or the motion vector field in MPEG-1/2.

Figure 3.27 shows the scope of the MPEG-7 standard. Note that feature extraction is outside the scope of MPEG-7, so is the search engine. This is similar to MPEG-1, which does not specify motion estimation, and MPEG-4, which does not specify segmentation algorithms. This approach is typical of most of the standard activities, i.e., "standardize the minimum". After MPEG-7 is finalized, various analysis tools, such as those mentioned in this chapter, can be further improved over time. This also leaves room for competition among vendors and researchers. Likewise, the query process (the search engine) should not be standardized, so that the design of search

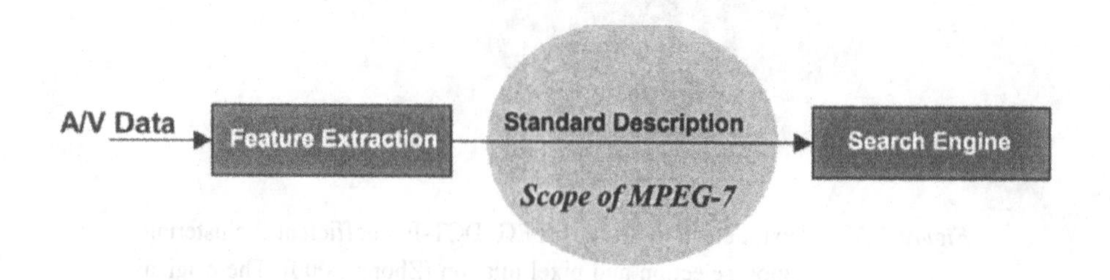

Figure 3.27 Illustration of the goal of the MPEG-7 standard.

engines and query languages to adapt to different application domains. MPEG-7 standardizes only what is necessary so that the description for the same content may adapt to different users and different application domains.

We now explain a few concepts of MPEG-7. One goal of the MPEG-7 is to provide a standardized method of describing features of multimedia data. For images and video, colors and motion are example features that are desirable in many applications. MPEG-7 defines a certain set of descriptors to describe these features. For example, the color histogram can be a very suitable descriptor for color characteristics of an image, and motion vectors (as commonly available in compressed video bit streams) form a useful descriptor for motion characteristics of a video clip.

MPEG-7 also uses the concept of description scheme (DS), which means a framework that defines the descriptors and their relationships. Hence, the descriptors are the basis of a description scheme. Description then implies an instantiation of a description scheme. MPEG-7 not only wants to standardize the description, it also wants the description to be efficient. Therefore, MPEG-7 also considers compression techniques to turn descriptions into coded descriptions. Compression reduces the amount of data that is stored or processed. Finally, MPEG-7 defines a description definition language (DDL) that can be used to define, modify, or combine descriptors and description schemes. Summarizing, MPEG-7 standardizes a set of descriptors and DS's, a DDL, and methods for coding the descriptions.

The process to define MPEG-7 was similar to those of the previous MPEG standards. Since 1996, the group has been working on defining and refining the requirements of MPEG-7, i.e., what MPEG-7 should provide. The MPEG-7 process includes a competitive phase followed by a collaborative phase. During the competitive phase, a Call for Proposals was issued and participants respond by both submitting written proposals and demonstrating the proposed techniques. Proposals were then evaluated by experts in video compression to determine merit. During the collaborative phase, MPEG-7 evolved as a series of experimentation models (XM), where each model outperforms the previous one.

Now finalized, MPEG-7 has a large variety of applications, such as digital libraries, multimedia directory services, broadcast media selection, multimedia authoring. Here are some examples. With MPEG-7, the user can draw a few lines on a screen to retrieve a set of images containing similar graphics. The user can also describe movements and relations between a number of objects to retrieve a list of video clips containing these objects with the described temporal and spatial relations. Also, for a given content, the user can describe actions and then get a list of scenarios where similar.

Summarizing, we have presented an overview of recent MPEG-7 activities and their strong relationship with audio, image and video characterization. Additional information on MPEG-7 may be found at the official website [MPEG 2003].

3.6 Embedded Content Features

A content feature is derived for describing actual content in an audio signal, transcript, image or video. In addition to the structure of the video, the motion, text and object features represent the actual content of image or video. Rather than interpreting text through language understanding, this book focuses on low-level audio and language features such as signal amplitude, speech discrimination, and keyword or keyphrase extraction.

Another form of content feature data is found in the procedures and methods for creating professional video. As described in chapter 2, video production manuals provide insight into the procedures used for video editing and creation. Pryluck wrote one of the widely referenced articles on video editing standards for the Journal of Broadcasting [Pryluck 1982].

A common and important element in video production is the ability to convey climax or suspense. Producers use a variety of effects ranging from camera positioning, lighting, and special effects for that purpose. Complete detection of these effects is beyond the realm of present image and language understanding technologies. However, many of the features described in sections 3.2, 3.3, and 3.4 are detectable or derivable in video.

Description Information

Descriptive video information is widely available in many production venues (chapter 2 discussed some of them). Simple features such as genre (documentaries, news footage, movies and sports) and duration may offer suggestions to assist in object recognition. For example, in news footage, the anchorperson will usually appear in the same pose and background at different times. The exact locations of the anchorperson can then be used to delineate story breaks. In documentaries, a person of expertise will appear at various points throughout the story when topical changes take place.

There are also many visual effects introduced during video editing and creation that may provide information for video content. For example, in documentaries the scenes prior to the introduction of a person usually describe their accomplishments and often precede scenes with large views of the person's face.

A producer will often create production notes that describe in detail the action and scenery of a video at every shot. If a particular feature is needed for an application in image or video databases, the description may have already been documented during video production.

GPS and Location Data

Another source of descriptive information is embedded with the video stream in the form of time code and geo-spatial (GPS/GIS) data. These features are useful in indexing precise segments in video or a particular location in spatial coordinates. Aeronautic and automobile surveillance video will often contain GPS data that may be used as a source for indexing. An example of geo-spatial video indexing and summarization is described in chapter 5.

3.7 Conclusions

This chapter provides an overview of automated and manual characterization methods of video. Sections 3.1 and 3.2 introduced the temporal representation and feature descriptions. Sections 3.3 and 3.4 described a number of audio, language, image and video features. Section 3.5 presented how many of these features are computed or approximated using encoded parameters in image and video compression. Section 3.6 was an overview of manual and embedded features.

Chapter 4 uses the results of these methods to interpret video structure and events. The events provide a means to prioritize video for summarization and characterization applications.

3.8 Bibliography

Akutsu, A. and Tonomura, Y. "Video Tomography: An efficient method for Camerawork Extraction and Motion Analysis," *Proceedings of the ACM International Multimedia Conference*, Oct., 1994, San Francisco, CA, pp. 349-356.

Arman, F., Depommier, R., Hsu, A., and Chiu, M. Y. "Content-based browsing of video sequences," *Proceedings of the ACM International Multimedia Conference*, October, 1994, San Francisco, CA pp. 97-103.

Arman, F., Hsu, A., and Chiu, M-Y. "Image Processing on Encoded Video Sequences," *Multimedia Systems* 1994 1, pp. 211-219.

Barron, J.L., Fleet, D.J., Beauchemin, S.S., "Performance of Optical Flow Techniques," *International Journal of Computer Vision,* 1994 12(1):43-77.

Biometric Consortium Notes, www.biometrics.org, *2003.*

Boccignone, G., De Santo, M., Percannella. G., "Automated Threshold Selection for the Detection of Dissolves in MPEG Videos". *IEEE International Conference on Multimedia and Expo (ICME)*, July 2000.

Bovik, A. "Handbook of Image and Video Processing," Academic Press Publishers, 2000.

Cohen, S. "Finding Color and Shape Patterns in Images", Technical Report and Ph.D. Thesis. STAN-CS-TR-99-1620, May 1999.

Deng, Y., Manjunath, B.S., C. Kenney, M.S.Moore, H.Shin, " An efficient color representation for image retrieval", *IEEE Transactions on Image Processing*, vol.10, (no.1), Jan. 2001. p.140-7.

Fablet, R., P. Bouthemy, P. Pérez, Non-parametric motion characterization using causal probabilistic models for video indexing and retrieval, *IEEE Transactions on Image Processing*, 11(4):393-407, April 2002.

Forsyth, D.A., Haddon, J., and S. Ioffe, "Finding objects by grouping primitives," in Shape, Contour and Grouping in Computer Vision, D.A. Forsyth, J.L. Mundy, R. Cipolla and V. DiGes'u (ed.s), Springer-Verlag LNCS 1681, 2000.

Furht, B., editor, "Multimedia Tools and Applications", Kluwer Academic Publisher, Norwell, MA, 1996.

Furht, B., Editor-in-Chief, "Handbook of Multimedia Computing", CRC Press, Boca Raton, Florida, 1999.

Gargi, Ullas, Kasturi, Rangachar , Strayer. Susan H., "Performance Characterization of Video-Shot-Change Detection Methods". *IEEE Transactions on Circuits and Systems for Video Technology*, Vol. 10, No. 1, Feb 2000.

Gong, Y., *et al*, " An Image Database System With Content Capturing and Fast Image Indexing Abilities", *IEEE International Conference on Multimedia Computing and Systems*, May, 1994.

Gonzalez, R. C. and Woods, R. E. "Digital Image Processing". New York: Addison-Wesley, 1992.

Hampapur, A., Jain, R., and Weymouth, T. "Production Model Based Digital Video Segmentation". *Multimedia Tools and Applications*, 1 (March 1995), 9-46.

Hauptmann, A.G., Speech Recognition in the Informedia Digital Video Library: Uses and Limitations, *ICTAI-95 7th IEEE International Conference on Tools with AI*, Washington, DC, November 6-8, 1995.

Hauptmann, A., Smith, M. "Text, Speech, and Vision for Video Segmentation," *AAAI Fall 1995 Symposium on Computational Models for Integrating Language and Vision.*

Hwang, M., Rosenfeld, R., Thayer, E., Mosur, R., Chase, L., Weide, R., Huang, X., Alleva, F., "Improving Speech Recognition Performance via Phone-Dependent VQ Codebooks and Adaptive Language Models in SPHINX-II." *ICASSP-1994*, vol. I, pp. 549-552.

Iqbal, Q., and Aggarwal, J. K., "Combining Structure, Color and Texture for Image Retrieval: A Performance Evaluation", *International Conference on Pattern Recognition*, August 2002, pp. 438-443.

Iqbal Q., and J. K. Aggarwal, J.K., "Perceptual Grouping for Image Retrieval and Classification", *3rd IEEE Computer Society Workshop on Perceptual Organization in Computer Vision*, July 8, 2001, Vancouver, Canada.

Jain, A. K., "Fundamentals of Digital Image Processing". Prentice-Hall, Inc., Englewood Cliffs, first edition, 1989.

Li, F., Doermann, D., and Kia, O., "Automatic Text Detection and Tracking in Digital Video". *IEEE Transactions on Image Processing.* Vol. 9, No. 1, pp. 147-156, Jan. 2000

Lienhart, R., "Reliable Transition Detection In Videos: A Survey and Practitioner's Guide". MRL technical report; International Journal of Image and Graphics (IJIG), August 2001.

Lienhart. R., "Comparison of Automatic Shot Boundary Detection Algorithms". *Storage and Retrieval for Still Image and Video Databases* VII 1999, Proc. SPIE 3656-29, January. 1999.

Lucas, B.D., Kanade, T. "An Iterative Technique of Image Registration and Its Application to Stereo," *Proceedings of the 7th International Joint Conference on Artificial Intelligence*, pp. 674-679, Aug. 1981.

Manjunath, B.S., J. -R. Ohm, V. V. Vinod, and A. Yamada, "Color and Texture descriptors," *IEEE Transactions on Circuits and Systems for Video Technology*, Special Issue on MPEG-7, 2001.

Mauldin, M. "Information Retrieval by Text Skimming," PhD Thesis, Carnegie Mellon University. August 1989. Revised edition published as "Conceptual Information Retrieval: A Case Study in Adaptive Partial Parsing, Kluwer Press, September 1991.

Meng, J., and Chang, S. F., "Tools for Compressed-Domain Video Indexing and Editing," *SPIE Conference on Storage and Retrieval for Image and Video Database*, San Jose, Feb. 1996.

"MPEG Digital Video Standard", Motion Picture Experts Group 2003, http://mpeg.telecomitalialab.com/.

Nam, Jeho, and Ahmed H. Tewfik, A. H., "Dissolve Transition Detection Using B-Splines Interpolation". *IEEE International Conference on Multimedia and Expo (ICME)*, July 2000.

Pryluck, C., "Meaning in Film/Video: Order, Time, and Ambiguity," JOURNAL OF BROADCASTING, 26(Summer 1982)3:685- 695 (with Charles Teddlie, Richard Sands).

Rowley, H., Kanade, T., and Baluja, S., "Neural Network-Based Face Detection ", *IEEE Transactions on Pattern Analysis and Machine Intelligence*, January 1998.

Rubner. Y., "Texture Metrics ". Phd Thesis, Stanford University, May 1999.

Rubner, Y., C. Tomasi, C., and Guibas, L. J. "A Metric for Distributions with Applications to Image Databases". Proceedings of the 1998 *IEEE International Conference on Computer Vision, ICCV*, Bombay, India, January 1998, pp. 59-66.

Salton, G., and McGill, M.J. "Introduction to Modern Information Retrieval," McGraw-Hill, New York, McGraw-Hill Computer Science Series, 1983.

Sato, T. Kanade, T., Hughes, E. Smith, M. "Video OCR for Digital News Archive," *IEEE International Workshop on Content-based Access of Image and Video Databases, ICCV,* Bombay, India, January 1998.

Schneiderman, H., and Kanade, T., "Object Detection Using the Statistics of Parts ", *International Journal of Computer Vision*, 2002.

Schneiderman, H., and Kanade, T.. "Probabilistic Modeling of Local Appearance and Spatial Relationships for Object Recognition." *IEEE Conference on Computer Vision and Pattern Recognition* (CVPR), pp. 45-51. 1998. Santa Barbara, CA.

Seidenberg, M., Bruce, B., Rubin, A., "Television Captioning for the Hearing-Impaired: A Case Study in Text Adaptation," BBN Report No. 5746, June 1981.

Shapiro, L.G. and A. Rosenfeld, "Computer Vision and Image Processing". Boston: Academic Press, 1992, edited volume.

Siegler, M., Jin, R. and Hauptmann, A., CMU Spoken Document Retrieval in TREC-8: Analysis of the role of Term Frequency TF, Proceedings of TREC-8, The eighth Text REtrieval Conference, NIST, Gaithersburg, MD, November 1999.Sleator, D., and Temperley, D., "Parsing English with a Link Grammar," *Third International Workshop on Parsing Technologies,* 1993.

Smith, M., Kanade, T. "Video Skimming and Characterization through the Combination of Image and Language Understanding Techniques". *Computer Vision and Pattern Recognition*. San Juan, PR, June 1997.

Smith. M., "Integration of Image, Audio, and Language Technology for Video Characterization and Variable-Rate Skimming", PhD Thesis, Department of Electrical and Computer Engineering, Carnegie Mellon University. January 1998.

Smith, M., and Chen, T. "Image and Video Indexing and Retrieval", Book Chapter, Handbook of Image and Video Processing," Al Bovik, Editor, Academic Press Publishers, Boston, MA 2000.

Sung, K-K, and Poggio, T., "Example-Based Learning for View-Based Human Face Detection", *Pattern Analysis and Machine Intelligence"*, January 1998, Vol 20. No. 1.

Tomasi, C., and Kanade, T. "Shape and Motion without Depth," *IEEE International Conference on Computer Vision, ICCV*, 1990, Osaka, Japan.

"TREC 93," *Proceedings of the Second Text Retrieval Conference*, D. Harmon, editor, sponsored by ARPA/SISTO, August 1993.

Tse, T.Y. and R. L. Baker, "Global Zoom/Pan Estimation and Compensation For Video Compression" *Proceedings of ICASSP 1991*, pp.2725-2728.

Wang, H., and Chang, S. F.,"A Highly Efficient System for Automatic Face Region Detection in MPEG Video Sequences," *IEEE Transactions on Circuits and Systems for Video Technology*, special issue on Multimedia Systems and Technologies, 1997.

Yeo, B. L., and B. Liu, B., "Rapid Scene Analysis on Compressed Video". *IEEE Transactions on Circuits and Systems for Video Technology*, Vol. 5, No. 6, pp. 533-544, December 1995.

Yeung, M., Yeo, B., Wolf, W., and Liu, B., "Video Browsing Using Clustering and Scene Transitions on Compressed Sequences". *Proceedings IS&T/SPIE Multimedia Computing and Networking*, February 1995.

Zabih, R., Miller, J., Mai, K., "A Feature-Based Algorithm for Detecting and Classifying Scene Breaks," *Proceedings of the ACM International Conference on Multimedia*, San Francisco, CA, November, 1995.

Zhang, H., Kankanhalli, A., and Smoliar, S. "Automatic partitioning of full-motion video," *Multimedia Systems* 1993 1, pp. 10-28.

Zhang, H.J., Tan, S., Smoliar, S., and Yihong, G."Video Parsing and Browsing Using Compressed Data," *Multimedia Tools and Applications*, 1, 1995.

Zhong, Yu, Zhang, Hongjiang and Jain, A.K.,"Automatic Caption Localization in Compressed Videos". *IEEE Transactions on Pattern Analysis and Machine Intelligence* (PAMI), Vol. 22, Issue 4, pp. 385-392, April 2000.

Chapter 4

Video Summarization

Video summarization selects appropriate keywords or keyphrases and a corresponding set of images to create a shortened video abstraction or video skim. This chapter describes automated methods for creating video summaries from multimodal features, in particular a rule-based method. In developing rules for selecting audio and image regions to be used in summaries, we describe the evolution of the video skim into other video surrogates in sections 4.1 and 4.2. The techniques in audio, language and image understanding provide a basis for automated video characterization. Audio characterization requires semantic understanding of the entire video, a task difficult even for most humans. Image characterization is limited to the detection of the characteristics as discussed in section 3.3. Methods for translating low-level features into individual audio and image candidates to be used in the video skim are discussed in sections 4.3 and 4.4. Rules for combining image and audio candidates may be generated from professional production standards. This process is described in section 4.5 through a set of primitive rules that prioritize skim candidates. Sections 4.6, 4.7, and 4.8 describe the highest order rules for specific content and visualization.

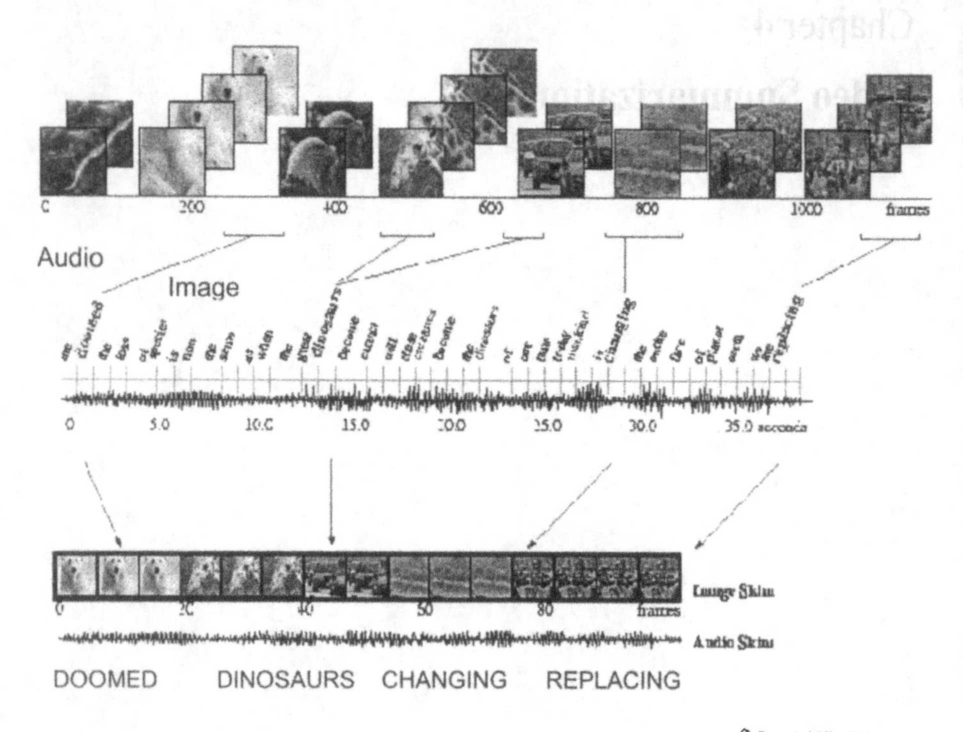

Figure 4.1 Keyword-based skim creation incorporating word relevance, significant objects (humans and text), and camera motion [Smith 1995]. An image region is assigned to each keyword: e.g., DOOMED - For this word the images following the camera motion are selected, DINOSAURS - The segment for "dinosaur" is long so portions of the next shot are used for more content, CHANGING - No significant image region for the word "changing", REPLACING - For this word, the latter portion of the shot contains both text and humans. This version of the skim appears like a slideshow during playback.

4.1 Video Skims

We start this chapter with the definition of Video Skim [Smith 1997] and [Smith 1998]. A skim is a compilation of "important" audio and image regions from a video into a smaller semantic unit. It results in a motion video that is much shorter than the original and retains the same semantic meaning. The first skim was based on individual keywords and short image sequences that resembled a slideshow [Smith 1995]. Figure 4.1 illustrates the concept of extracting separate audio keywords and image regions to form the video skim.

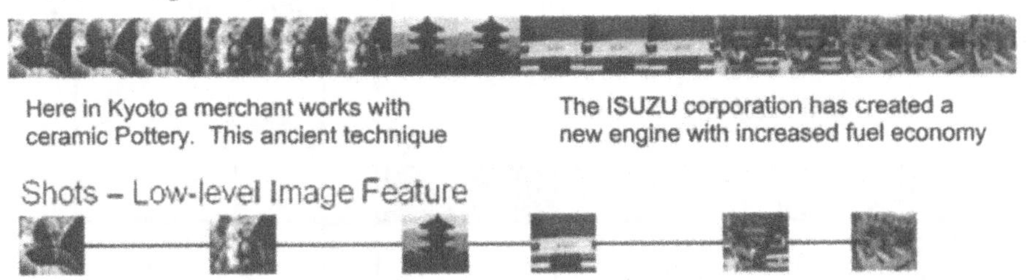
Video - Image and Audio
Here in Kyoto a merchant works with ceramic Pottery. This ancient technique
The ISUZU corporation has created a new engine with increased fuel economy
Shots – Low-level Image Feature
Phrase – Mid-level Audio Feature
Here in Kyoto a merchant works with ceramic Pottery. This ancient technique
The ISUZU corporation has created a new engine with increased fuel economy
Segments – High-level Application

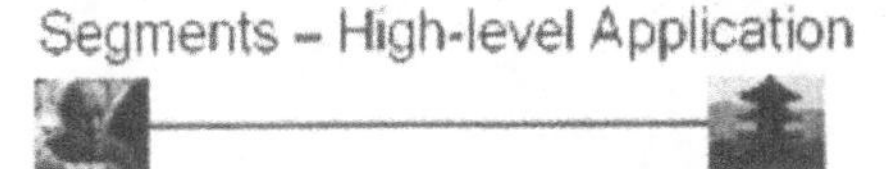
Here in Kyoto a merchant works with ceramic Pottery. This ancient technique

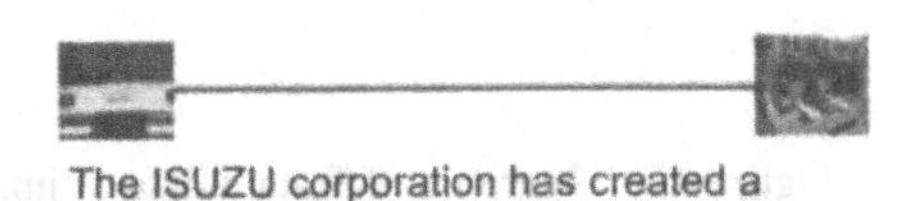
The ISUZU corporation has created a new engine with increased fuel economy

Figure 4.2 Illustration of three-tier video segmentation. The low-level feature is a shot change. The mid-level feature is an audio keyphrase and the high-level is the separation of the story into two segments. The story is about advances in ceramics, with the first segment on ancient methods and the second segment on modern ceramics.

The integration of multiple modalities is the foundation for video characterization and skimming. By integrating features, we can better interpret video content for meaningful visualization applications. Figure 4.2 illustrates this concept with a three-tier level of feature analysis. Low-level features describe statistical content; mid-level features use low-level features to interpret video content; and high-level features describe a type of output – in this case video segmentation. The three-tier description is most appropriate in video characterization as it outlines traditional features, content analysis and applications. In general, the number of tiers may be as numerous as the desired level of analysis. The sections below describe the steps involved in selecting, prioritizing and ordering keywords and video frames for skims and other forms of summarization and segmentation.

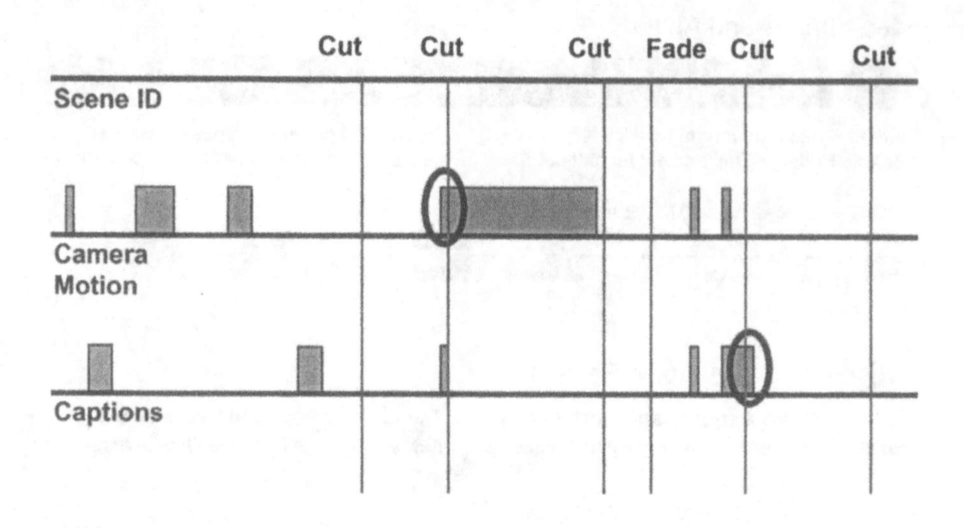

Figure 4.3 Merging of feature data to improve boundary accuracy. Detected motion and text captions extend beyond a shot cut prior to pruning.

Feature Correction

One application for multimodal integration is the merging of image and audio metadata for improving the accuracy of individual features. Many of the feature extraction methods in chapter 3 use a windowing operation to estimate a feature over time. In motion detection, for example, the results of analyzing an individual frame are averaged over a 20 to 30-frame window. This provides a more accurate assessment of the type of motion. The same process is used in face detection, text detection and audio parsing.

In the case of the shot change, the accuracy is more granular, and its shot boundaries may be used to prune or extend the boundaries of other features. An example of this is illustrated in Figure 4.3, where shot changes provide a basis for correcting motion estimation boundaries. In most cases, motion will not end abruptly or extend beyond a shot change. The motion boundaries are pruned or extended as necessary to fit within a shot.

4.2 Automatic Video Surrogates

The concept of summarization has existed for some time in areas such as text abstraction, video editing, image storyboards and other applications. In Firmin's evaluation of automatic text summarization systems, he divides summaries into the following three categories: 1.) Indicative, providing an indication of a central topic, 2.) Informative, summaries that serve as substitutes for full documents, and 3.) Evaluative, summaries that express the author's point of view [Firmin 1999]. For video, the function of a summary is similar, but there are additional opportunities to express the result as text, imagery, audio, video or some combination. Video summaries may also appear very different from one application to the next. Just as summaries can serve different purposes, their composition is greatly determined by the video genre being presented [Li 2000]. For sports, visuals contain lots of information; for interviews, shot changes and imagery provide little information; and for classroom lecture, audio is very important.

Summarization is inherently difficult because a perfect job requires complete semantic understanding of the entire video (a task difficult for the average human). The best image understanding algorithms at the moment can only detect simple characteristics like those discussed in chapter 3. In the audio track we have keywords and segment boundaries. In the image track we have shot breaks, camera motion, object motion, text captions, human faces and other static image properties.

Multimodal Processing for Deriving Video Surrogates

A video can be represented with a single thumbnail image. Such a single image surrogate is not likely to be a generic, indicative summary for the video in the news and documentary genres. Rather, the single image can serve as a query-relevant indicative summary. News and documentary videos have visual richness that would be difficult to capture with a single extracted image. For example, viewers interested in a NASA video about Apollo 11 may be interested in the liftoff, experiences of the astronauts on the moon, or an expert retrospective looking back on the significance of that mission. Based on the viewer's query, the thumbnail image could then be a shot of the rocket blasting off, the astronauts walking on the moon, or a head shot of the expert discussing the mission, respectively. Figure 4.4 illustrates how a single representative thumbnail could be chosen from a group of shot key frame images.

Given the NASA video, speech recognition breaks the dialogue into time-aligned sequences of words. Image processing breaks the visuals down into a sequence of time-aligned shots. Further image processing then extracts a single image, i.e., a frame, from the video to represent each shot. When the user issues a query, "walk on the moon", for example, language processing can isolate the query terms to emphasize ("walk" and "moon"), derive additional forms ("walking" for "walk"), and identify matching terms in the dialogue text. In the case illustrated in Figure 4.4, the query "walk on the moon" matches most often to the shot at time 3:09 when the dialogue includes the phrase "Walking on the moon." The shot at time 3:09 is represented by a black and white shot of two astronauts on the moon, and the key frame image of this shot is then chosen as the image surrogate for this video. Speech, image, and language processing each contribute to the derivation of a single image, which represents a video document.

4.3 Audio Skim Selection

Another first level of analysis for summarization is the creation of the reduced audio track, which is based on the keywords. Keywords are selected based on audio and language characterization. An example of extracting keywords from the original segment is shown in Figure 4.5. To reduce fragmented audio and increase comprehension, longer audio sequences are selected for the summary.

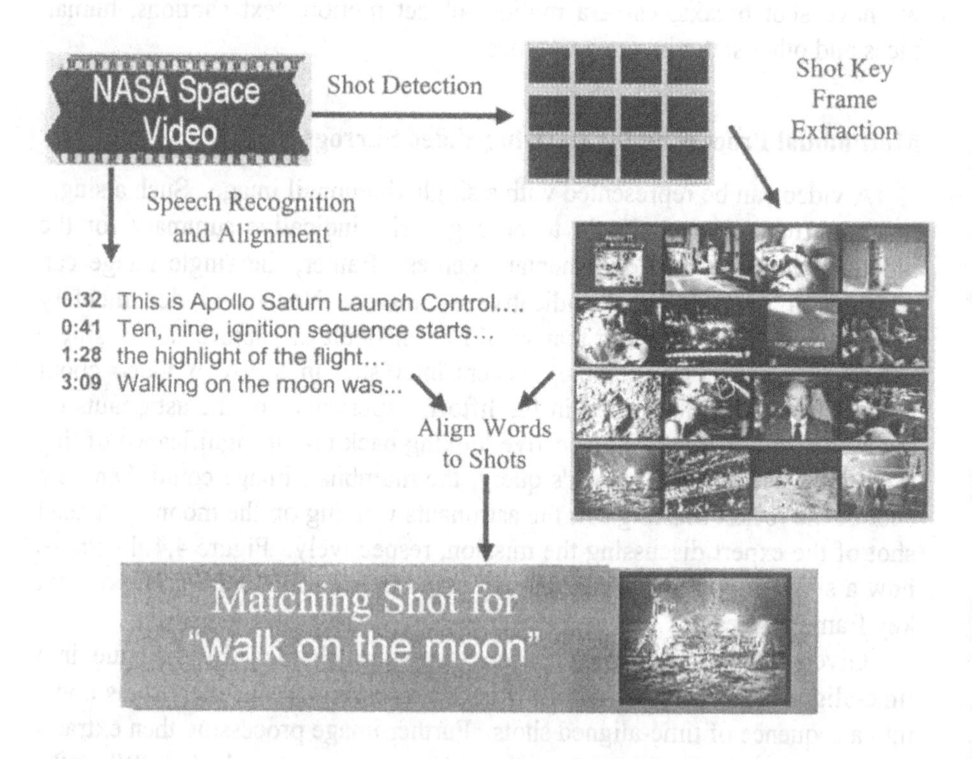

Figure 4.4 Thumbnail selection from shot key frames aligned with words. Thumbnails align with a query word in video retrieval. The selected image frame has greater correspondence with the topic of interest.

Background

The amount of legible or pertinent speech in many video sequences is quite small. This low density of relevant speech is one factor that allows regional audio exclusion for video summaries. The audio samples in this section were encoded at 44Khz in the MPEG-1 audio compression standard [MPEG 2003]. The audio was subsampled to 8Khz and converted to a raw data file for processing.

Keywords

TF-IDF weighting is the primary method for determining word relevance. It is adjusted to emphasize certain words. Conjunction words receive twice their TF-IDF value and question words receive three times their TF-IDF value during keyword or keyphrase detection. A "slang" word will receive a higher weight only if the source video contains relatively few "slang" words. The TF-IDF value for a proper noun does not change, but the selection priority for skimming will vary depending on the type of video.

One method of keyword selection is to choose the words whose TF-IDF values are higher than a fixed threshold. By varying this threshold, we control the number of keywords, and thus, the length of the skim. Keyword selection is straightforward in the ASCII transcript, but extracting these words from the audio track is difficult.

A user specified compaction level determines the length of the summarized audio. The actual length of the new audio track is not exact due to variations in duration of the keywords. Figure 4.5 illustrates an example of extracting keywords for summary creation. We will see later, however, that concatenation of extracted single keywords produces a summary that is audibly unintelligible for most users, and that keyphrases need to be extracted, instead. The TF-IDF values may also be used to weight these keyphrases.

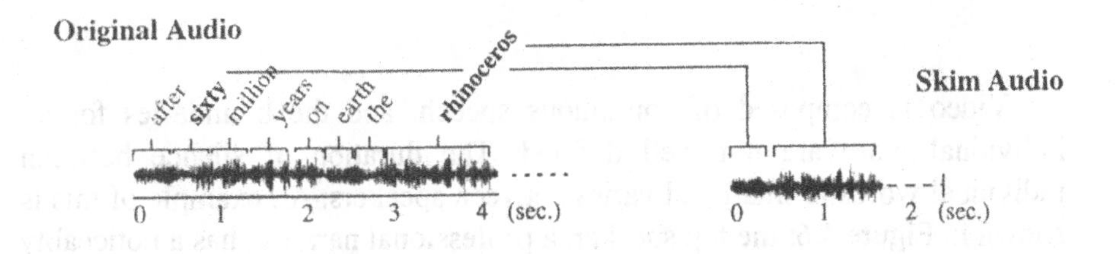

Figure 4.5 The word length for "rhinoceros" is 1.10 seconds, which allows for 33 frames; the word "sixty" uses 19 video frames or 0.63 seconds of audio. Keywords are added until the audio skim length is filled.

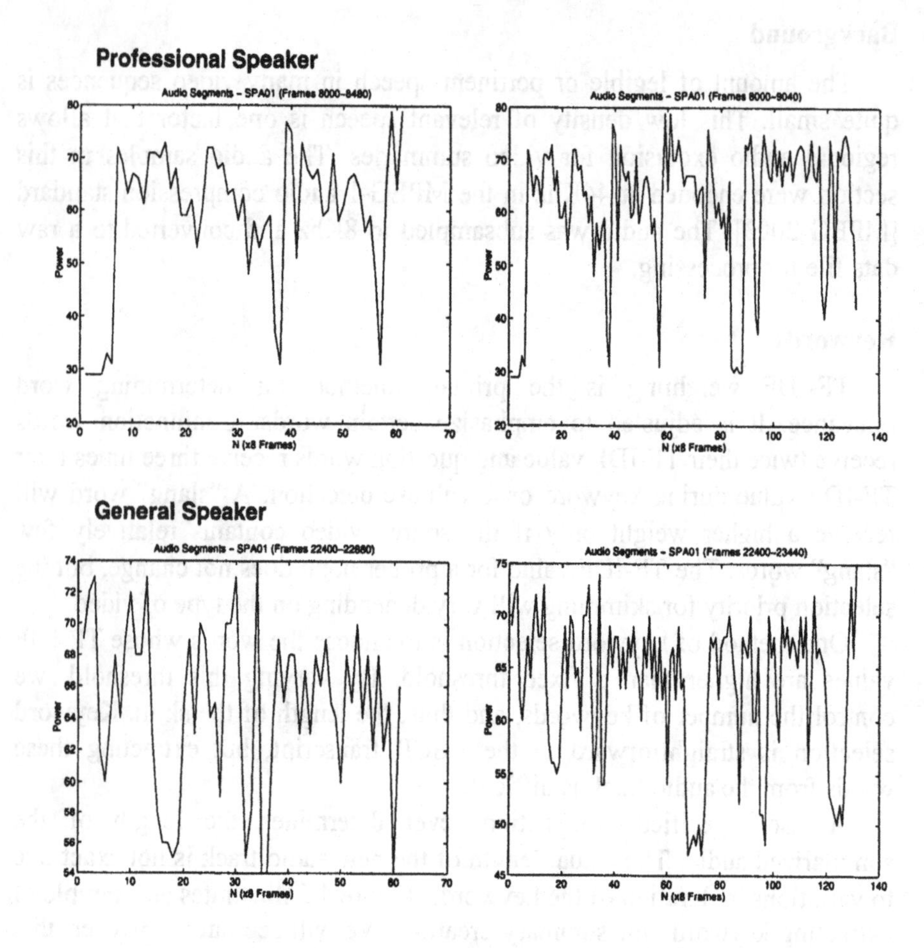

Figure 4.6 Varying levels of resolution for speech patterns: Top) Professional Female Anchorperson – Opening for a broadcast news segment. Bottom) Interviewee – Random male subject speaking over loud background noise. The audio signal of the anchorperson is clearer and easier to recognize with speech recognition than the signal of a random speaker.

Video is composed of continuous speech, and the boundaries for an individual word are not well defined. The duration of silence between individual words is small and varies between speakers. An example of this is shown in Figure 4.6; the top speaker, a professional narrator, has a noticeably clear speech signal, while the bottom speaker, an interviewee in a laboratory, has a noisier speech signal. The lower bound for the narrator power signal is roughly 30 db, while the power signal for the general speaker has a lower bound of roughly 50 db. The quality of speech recognition changes dramatically between speaker and environment [Hauptmann 1995].

Professional narrators use close-talking studio environments to record audio. Interviews often take place in field locations with background noise.

Keywords that appear in close proximity or repeat throughout the transcript may create skims with redundant audio. Therefore, it is necessary to discard keywords that repeat within a minimum number of frames (300 frames) and limit the repetition of each word. Using individual keywords creates an audio skim that is fragmented and incomprehensible for some speakers [Smith 1998] and [Christel 1998]. The skim from Figure 4.1 uses single keywords for the reduced audio track.

Keyphrase Analysis

Another problem with selecting single keywords for summarization is the short duration of individual words. Most audio words are less than 0.3 seconds in duration, and therefore, very difficult to extract. The first skims were created with keywords and manual audio segmentation. When used successively in the skim, the pace of the audio is fast and difficult to understand. A keyphrase should contain legible audio with a grain-size that is long enough for interpretation by the human listener.

Figure 4.7 illustrates an example of a video summary with keyphrases. This is the same documentary from Figure 4.1, but the audio phrases are much longer. For example, the word "doomed" is replaced by the phrase, "many creatures are doomed".

- **Fixed Extensions**

  In order for skims to be legible, the grain-size for a single audio region must be greater than the length of a single word. A potential solution is to buffer the keywords with a small period of silence. Although this improves the pace of the skim audio, the content is still based on single words. Using more than just a single keyword or a keyphrase can increase the understanding of an audio segment and convey more content. One method of keyphrase selection is to simply extend the boundary of a keyword by a fixed amount. This will certainly create a longer audio segment, but the content is unknown, and the end-points may crop the ending word. The keyword boundaries may instead be extended to another keyword. In this case, the keyphrase is longer, but audio segmentation is still a problem. The end word is not cropped, but accurate segmentation is difficult, so there is likely word truncation. An extension of the keyword to a period of silence, $Sw < 0.2$ seconds, solves the problem of word truncation, but the content is questionable.

- **Phrase Parsing**

  The Link Grammar Parser [Sleator 1993] may be used to provide noun phrases as a tool for keyphrase extraction (Section 3.3). The sum of the TF-IDF values of the words is used to weight the keyphrases. With this method, the content is understood, but there is still the problem of audio segmentation. The words at a boundary of a keyphrase can be truncated, causing a jerky sound during playback. If we extend the keyphrase to a region of silence, we now have a content-based keyphrase with legible audio.

  Using a grammar parser has the same drawbacks in keyphrase detection as methods that use single keywords, keyword-to-keyword, and keyword-to-silence parsing. These methods assume the speech recognizer has accurate word alignment. Even with the grammar parser and extensions to silence, the keyphrases are often long and unintelligible, as shown in section 3.3.

- **Audio Segmentation**

  There are natural pauses in human speech, that when segmented properly, reflect the intended statements in the audio track. A more intuitive approach to keyphrase extraction is to use the boundaries from audio segmentation to select spoken phrases. Until now, these boundaries have been used to indicate silence between words. If we tune our parameters for longer pauses on the order of $S_w >= 0.5$ seconds, we will extract keyphrases with high correspondence to human speech patterns. Speech recognition is used to decipher the words, and the total TF-IDF value is used to weigh the keyphrases. With this method, we do not rely on speech recognition as the first pass in keyphrase extraction. In most cases, the resulting audio is legible. There are cases when this method fails to produce a perfect phrase. A dramatic pause will yield an incomplete statement, and often speech will continue for several seconds without a pause. In section 4.5, the parameters for the default and genre specific values for keyphrase audio segmentation are discussed.

  When speech recognition is errorful, the number of detected keywords is limited. This smaller set of words can be weighted by TF-IDF or a confidence measure from the speech recognition system. Segmented audio is weighted according to parsing accuracy and proximity to significant imagery, thus there is no dependence on speech recognition or word alignment. Speech phrases can also be distinguished from music or noise using a variety of methods [Kedem 1986], [Saunders 1996], and [Jarina 2001]. The average keyphrase grain-size is 7 seconds for documentary footage.

Video Coverage

TF-IDF values can be computed from the entire video or a smaller segment. Early experiments computed term-frequency with the entire video and not individual segments [Smith 1995]. This can produce a skim with audio regions clustered at different points in the video. To provide even coverage of the original segment, an audio keyphrase is chosen at fixed intervals. The interval for selection, I_w, is based on the compaction level, the duration of the original segment, and the granularity of spoken audio.

When the original segment is small (less than 2 minutes), the coverage is concentrated at the beginning and end of the segment. The opening and closing scenes tend to summarize the content of the whole video. In chapter 6, we discuss the effects of even and uneven coverage in the audio and image selection process. The compaction level will affect grain size, as described later in chapter 5. A user may specify important keywords that will affect the skim coverage. In video libraries, queries are based on keywords that provide additional preference for the regions containing these words in the video.

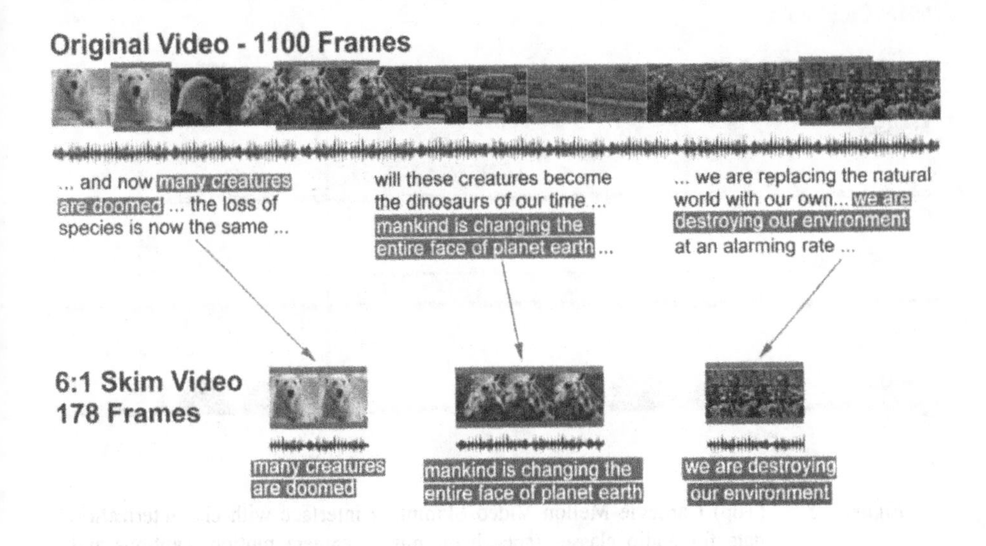
Original Video - 1100 Frames
... and now many creatures are doomed ... the loss of species is now the same ...
will these creatures become the dinosaurs of our time ... mankind is changing the entire face of planet earth ...
... we are replacing the natural world with our own... we are destroying our environment at an alarming rate ...
6:1 Skim Video
178 Frames
many creatures are doomed
mankind is changing the entire face of planet earth
we are destroying our environment

Figure 4.7 Skim creation incorporating audio keyphrases and significant images [Smith 1997]. The tempo of this skim closely resembles that of the original video during playback.

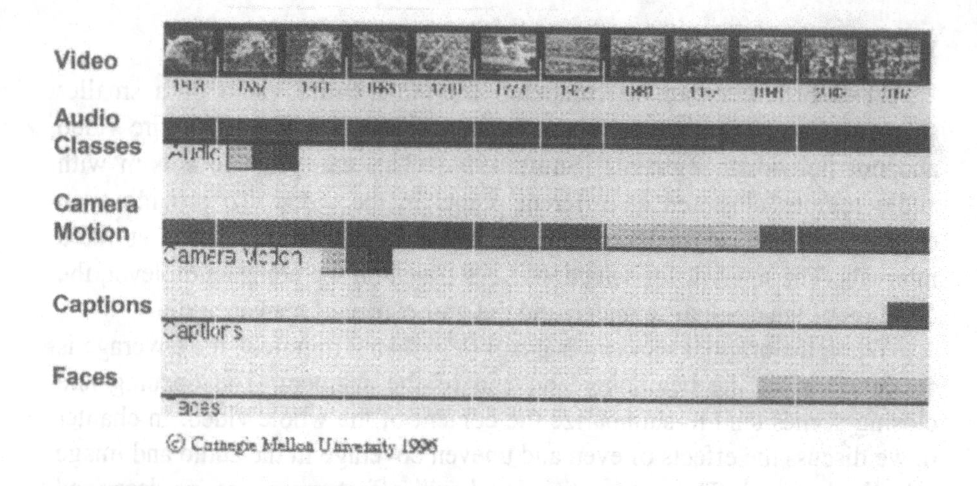

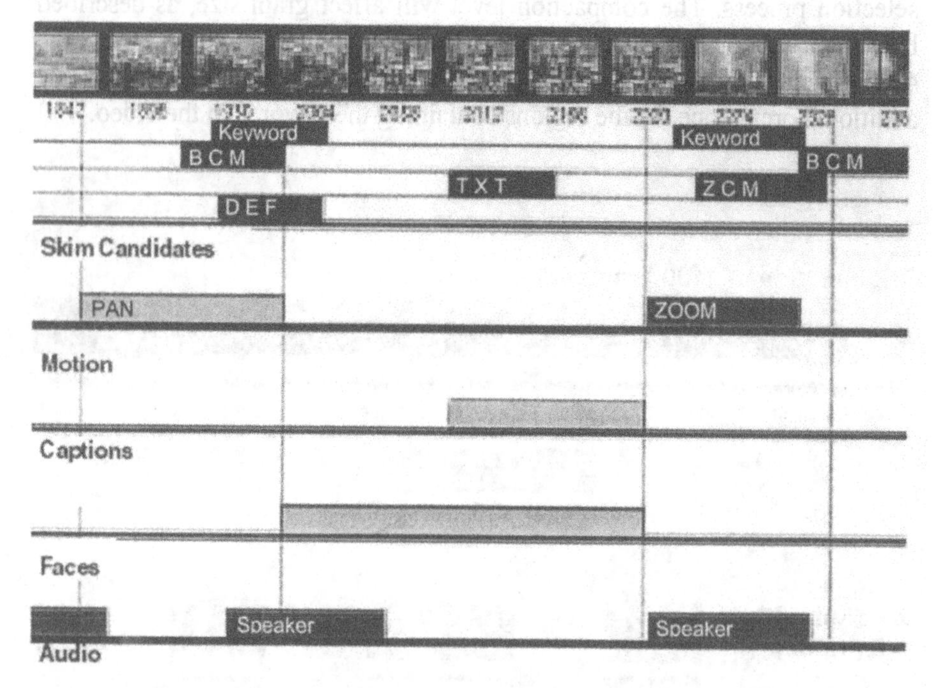

Figure 4.8 (Top) Carnegie Mellon Video Skimming interface with characterization data for audio classes (speech or music), camera motion, captions and human faces. (Bottom) Characterization data with skim candidates and keyphrases for "Destruction of Species". The skim candidate symbols correspond to the following primitive rules: BCM, Bounded Camera Motion; ZCM, Zoom Camera Motion; TXT, Text Captions; and DEF Default. Vertical lines represent shot breaks. The top temporal characterization graph displays metadata only. The bottom graph shows metadata and skim candidates with ranked keywords.

4.4 Image Skim Selection

After the audio track has been processed for keyword or keyphrase selection, an appropriate image track is chosen to best fit the criteria of the summarization. In order to create the image skim, we might think of selecting those video frames that correspond in time to the audio skim segments. As we often observe in television programs, however, the contents of the audio and video are not necessarily synchronized. There are also many visual effects introduced during video editing and creation that do not necessarily convey content. Therefore, for each keyword or keyphrase we must analyze the characterization results of the surrounding video frames and select a set of frames which may not align with the audio in time, but which are most appropriate for summarization.

A shot is the basic visual element of analysis in video understanding. For each keyphrase, we analyze the image characterization data for the corresponding shot, as well as, the surrounding shots. The number of surrounding shots, W_s, will vary according to the type of video. For most videos W_s is one shot back and one shot forward in time. To prevent using shots that do not appear in close proximity of the keyphrase, the temporal search space is limited by a fixed duration, W_t. The approximate value of W_t will also vary according to the video genre. The image and audio characterization data is analyzed together to select the image portion of the skim. Images that are not synchronized with the keyphrase may provide more information.

The characterization features can provide insights into the structure of video, but this data is only useful for specific rules. For some rules, the image characterization will control the selection of a skim region, rather than first selecting an audio region and then a corresponding image region.

4.5 Primitive Skim Selection Rules

By using examples from the manual skim study and video production standards, we identify an initial set of heuristic rules for automatically selecting skim video. The first heuristics are a set of primitive rules. They are applied to the video frames in the shot containing the keyword or keyphrase, and the shots that follow within the search window, W_t.

The development of rules that use images and audio is facilitated through a temporal characterization display. Figure 4.8 shows an example output of the characterization data with keywords or keyphrases and image skim candidates from the Carnegie Mellon Video Skimming system. For each keyword, the four rows above "Skim Candidates" indicate three candidate image sections selected by primitive rules and the keyword position. Note that each image candidate appears in the first or second shot of the keyword. Table 4.1 shows the video characterization data and the skim candidates selected with the primitive rules. These primitive rules (PR) are described below in order of priority. They are based on video phenomena or events that indicate certain video content.

PR-1 Keyphrase Near a Fade or Dissolve (KNF)

Fade shot changes or dissolves usually imply a change in content. When a keyphrase is in close proximity to this type of shot change, we use the image region immediately following the fade or dissolve. Shot change categories and detection methods are discussed in chapters 2 and 3 [Boreczky 1996], [Gargi 2000], and [Lienhart 2001].

PR-2 Keyphrase Preceding a Cut (KNC)

Shot changes, which are simple cuts, do not necessarily imply a change in content. However, they provide a useful means for identifying important image regions for skimming. An important statement will often precede a small pause and a shot change. When a keyphrase appears in close proximity (between 30 and 80 frames) to a shot cut, the image portion following the cut is used for skimming. If the keyphrase duration is long, a small portion (at least 60 frames) of the shot containing the keyphrase is included in the following shot, as discussed later in the visual clarity portion of section 4.8.

PR-3 Bounded Camera Motion and Zoom (BCM/ZCM)

Video producers often use camera motion as a tool to highlight a portion of a shot. The frames that precede or follow a pan or zoom motion are usually the focus of the segment. Knowing which frames exhibit camera motion, we can isolate the video regions that are static, and therefore likely to be the focal point in a scene. When static frames surround camera motion, the frames that follow usually contain more visual content. The video frames that precede or follow a pan or zoom motion are usually the focal point of the segment. We can isolate the video regions that are static and bounded by segments with motion, and therefore likely to be the focal point in a scene containing motion. Figure 4.9 shows example outputs of the Bounded Camera Motion and Zoom Motion Rule.

PR-4 Object Motion (OBM)

Object motion is important because video producers usually include moving subjects to show action. When moving objects are present, we use the beginning portion of the object motion as the start of image skim.

Table 4.1 Rules and Required Features

Primitive Rule (Acronym)	**Required Features**
PR-1 Keyword Near Fade (KNF)	• Keyword Detection • Shot Change Type
PR-2 Keyword Near Cut (KNC)	• Keyword Detection • Shot Changes
PR-3 Bounded Camera Motion (BCM) Zoom Shots (ZCM)	• Keyword Detection • Camera Motion • Partial Object Motion
PR-4 Significant Object Motion (OBM)	• Keyword Detection • Object and Camera Motion
PR-5 Video Captions (TXT)	• Keyword Detection • Text Detection
PR-6 Human Faces (FAC)	• Keyword Detection • Face Detection • Object Motion (moving heads)
PR-7 Significant Audio (AUD)	• Audio Classification • Shot Changes
PR-8 Default Rule (DEF)	• Keyword Detection • Object and Camera Motion • Shot Changes
Meta-Rule (Acronym)	**Required Primitive Rules and Features**
MR-1 Selection of Audio from Image Analysis	• All Primitive Rules and Features
MR-2 Introduction of Speaker (INS)	• Proper Noun Keyword • **PR-5** Video Captions (TXT) • **PR-6** Human Faces (FAC) • Image Similarity
MR-3 Inter-cuts or Image Adjacencies (SIS)	• **PR-2** Keyword Near Cut (KNC) • Image Similarity
MR-4 Successive Short Shots (SS1)	• Keyword Detection • **PR-2** Keyword Near Cut (KNC) • Shot Changes

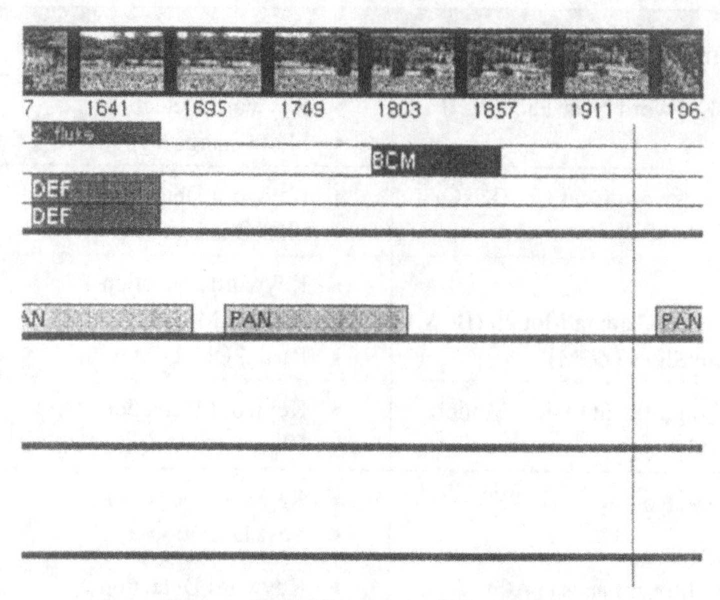

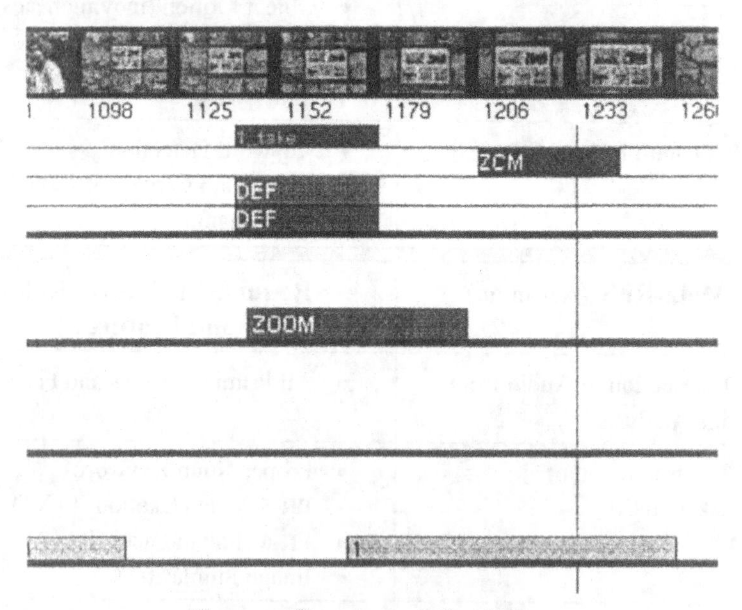

Figure 4.9 Bounded Camera Motion and Zoom Sequences detected for skim selection. Top) The images after the pan are selected. Bottom) The images at the end of the zoom are selected.

PR-5 Video Captions (TXT)

A shot will often contain captioned text to describe the content. Captions are used to indicate name and title, location, date, and other useful information. When they are present, the starting caption frame is used as the starting image frame for skimming. Depending on the type of video, the location of video captions will supersede a key-phrase in skim selection. Figure 4.10 is an example of the results from the Video Captions Rule.

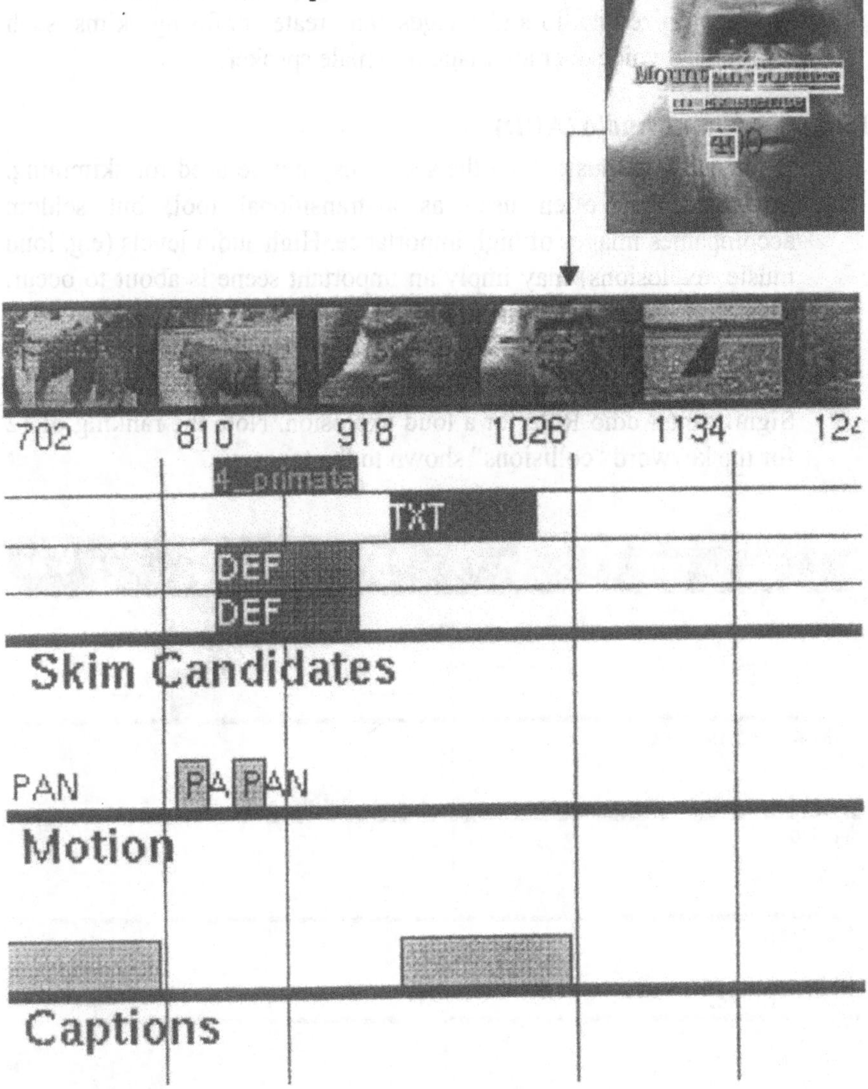

Figure 4.10 Experimental system that selects on-screen text regions for skimming. Note the number "4" as the ranking for the keyword "primate". Text detection for this image is described in Figure 3.19.

PR-6 Human Faces (FAC)

A shot containing human faces describes content and perspective associated with the human subject. The audio and image skim should be synchronized, even when video characterization selects different image and audio regions. When a "talking head" is speaking for a long period, audio selection does not change, but the skim rules search for important images from adjacent shots. Audio overlaid with faces from a different shot will appear disoriented. The unsynchronized audio and images can create confusing skims, such as a female voice over an image of a male speaker.

PR-7 Significant Audio (AUD)

If the audio is music, then the scene may not be used for skimming. Soft music is often used as a transitional tool, but seldom accompanies images of high importance. High audio levels (e.g. loud music, explosions) may imply an important scene is about to occur. The skim region will start after high audio levels or music. In most cases, we cannot segment speech accompanied by music. We set unclassified audio as music. Figure 4.11 is an example of the Significant Audio Rule for a loud explosion. Note the ranking of 12 for the keyword "collisions" shown in the top row.

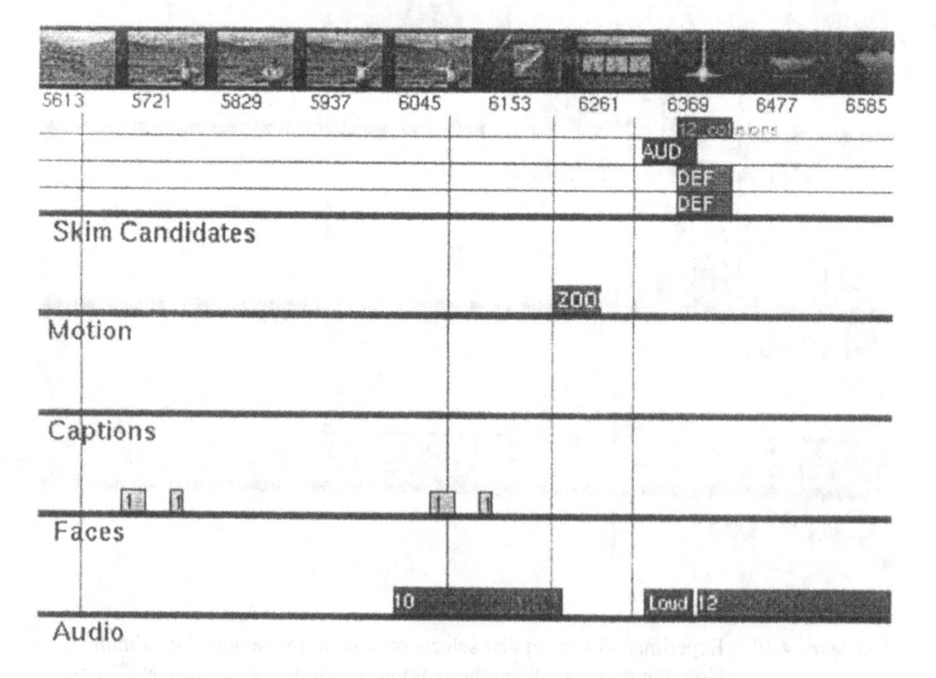

Figure 4.11 Significant Audio detected during a loud explosion.

PR-8 Default Rule (DEF)

Default video frames align to the audio keyphrases. The skim images are the frames of the shot that correspond to the keyword or keyphrase. Default rule selections are used as the second and third image candidate in Figures 4.9, 4.10 and 4.11. This rule is also the selection criteria for shots containing a large number of camera or object motion regions. An example is shown in Figure 4.15b, where random camera movement results in multiple pans and zooms. This excess motion often signifies action or live footage.

4.6 Feature Integration Meta-Rules

Summary creation involves selecting the appropriate keywords or audio and choosing a corresponding set of images. Primitive Rules, described in 4.5, are independent rules that provide candidates for the selection of image regions for a given keyword. We use another set of rules, Meta-Rules, as higher order rules that select a single candidate from the primitive rules according to global properties of the video. In some rules, image selection is more important and takes place before audio keyphrase selection.

The features described in previous sections may be used in rules that describe a particular type of video shot to create an additional set of content features, such as shot structure [Furht 1996], audio [Furht 1999], objects [Forsyth 2000], and face [Schneiderman 2002]. Video production standards are used to identify a small set of heuristic Meta-rules for assessing a "summary rank" or priority to a given subset of video. These rules involve mostly the integration of image processing features with audio and language features. The bottom of Table 4.1 shows the Primitive Rules and features used for Meta-rules. Below is a description of the Meta-rules (MR) suitable for most types of video.

MR-1 Selection of Audio from Image Analysis

The first meta-rule is the selection or ranking of audio based on image properties. Audio is parsed and recognized by speech recognition, keyword spotting and other language and audio analysis procedures. For this rule, keyword ranking is increased based on its proximity to certain imagery. For example, if an audio phrase with a high keyword value is within a short duration of a video caption, it will receive a higher summary rank than an audio phrase with a similar keyword value. An example of this rule is shown in Figure 4.12.

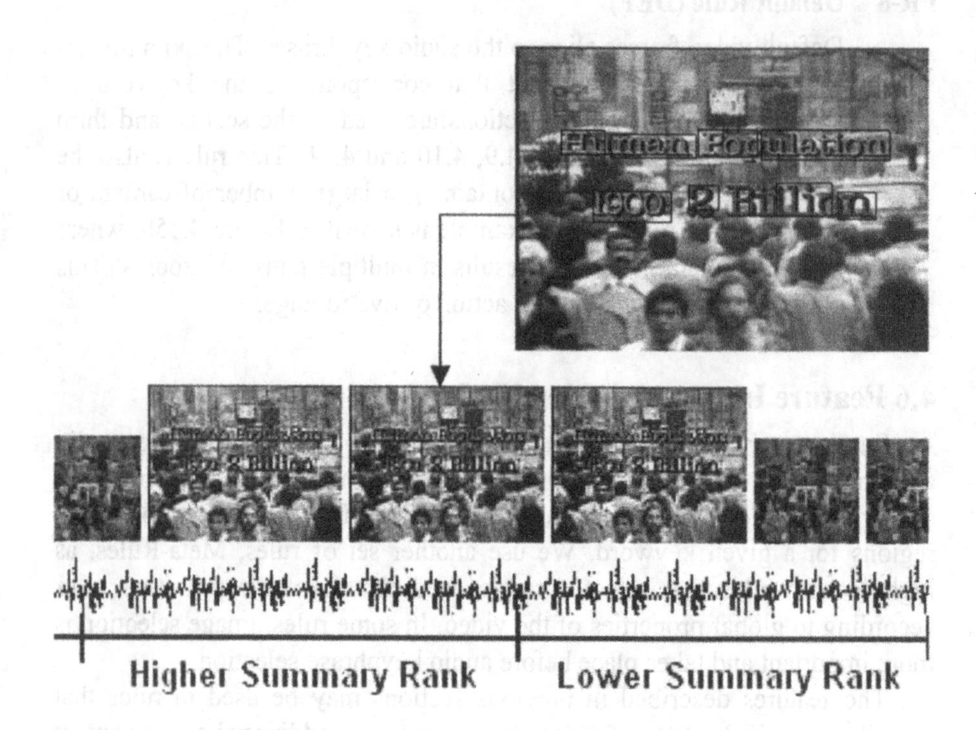

Figure 4.12 Audio with unsynchronised imagery. The lower ranking is given to audio near images with faces and no text. The audio in close proximity to the images with both a face and text receives a higher rank.

MR-2 Introduction Scenes (INS)

The shots prior to the introduction of a person usually describe their accomplishments and often precede shots with large views of the person's face. A person's name is spoken and then followed by supportive material. Afterwards, the person's actual face is shown. If a shot contains a proper name or an on-screen text title, and a large human face is detected in the shots that follow, we call this an Introduction Scene. Characterization of scenes of this type is useful when searching for a particular human subject because identification of the particular person is more reliable than using the image or audio features separately. The Introduction Rule is common in documentaries that focus on human achievement.

We may also identify occurrences of a person appearing later in the video. In most cases, the face detection will not produce an accurate match unless the face is in the same pose and position. The name may not be recognized with speech recognition, but closed-captions or a perfect transcript may be available as a second point of analysis. For this rule, we first locate a proper noun in the keyphrase and then look for a human face. An illustration of the Introduction Rule is shown in Figure 4.13. The System Output shows a result from the Carnegie Mellon system [Smith 1997].

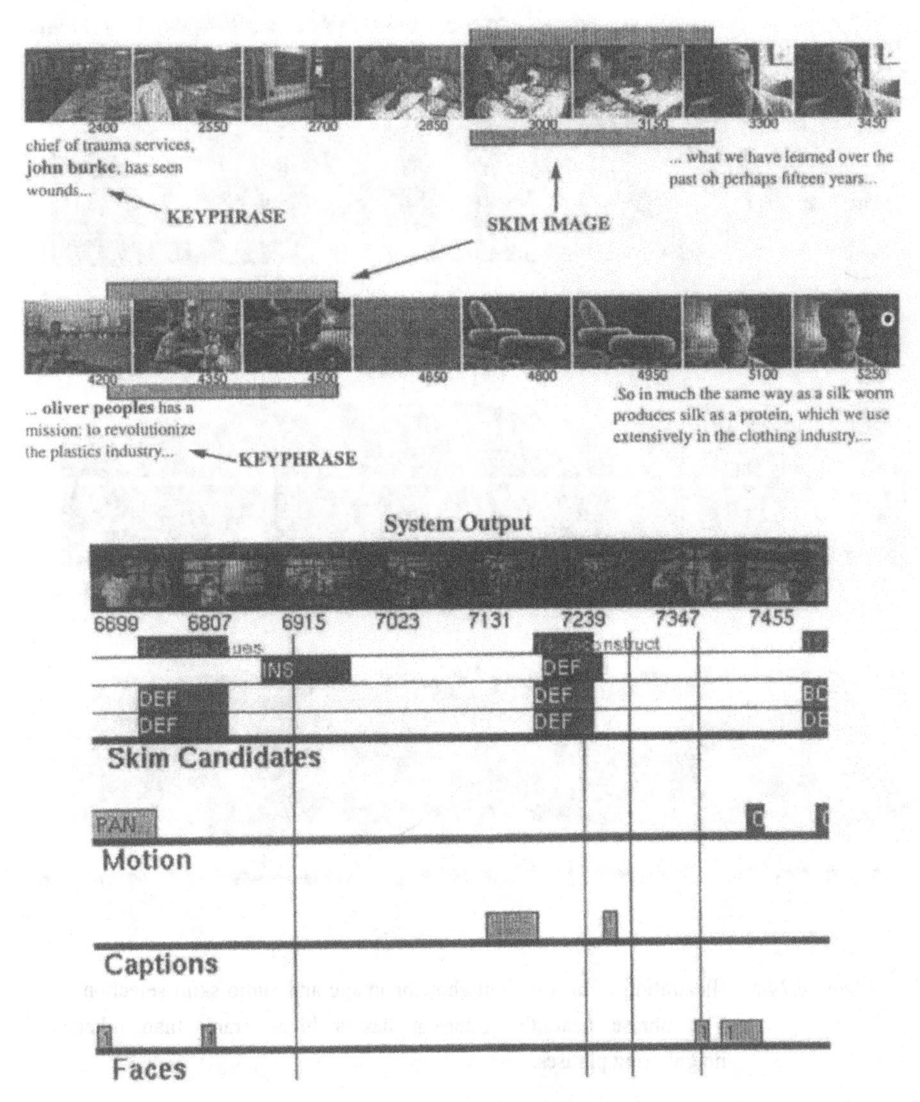

Figure 4.13 Two examples of identification of an "Introduction Scene". In both examples, a name is uttered followed by a supportive shot or scene, and then a close-up of a human face. Top) Illustration of the "introduction rule" with two scientists working on artificial skin, Bottom) Result from an experimental skimming system. The keyphrase ranked 13 contains the word "colleagues" and a proper noun. An introduction scene is chosen that overlaps the adjacent shot. The parameters for determining a facial close-up are empirical.

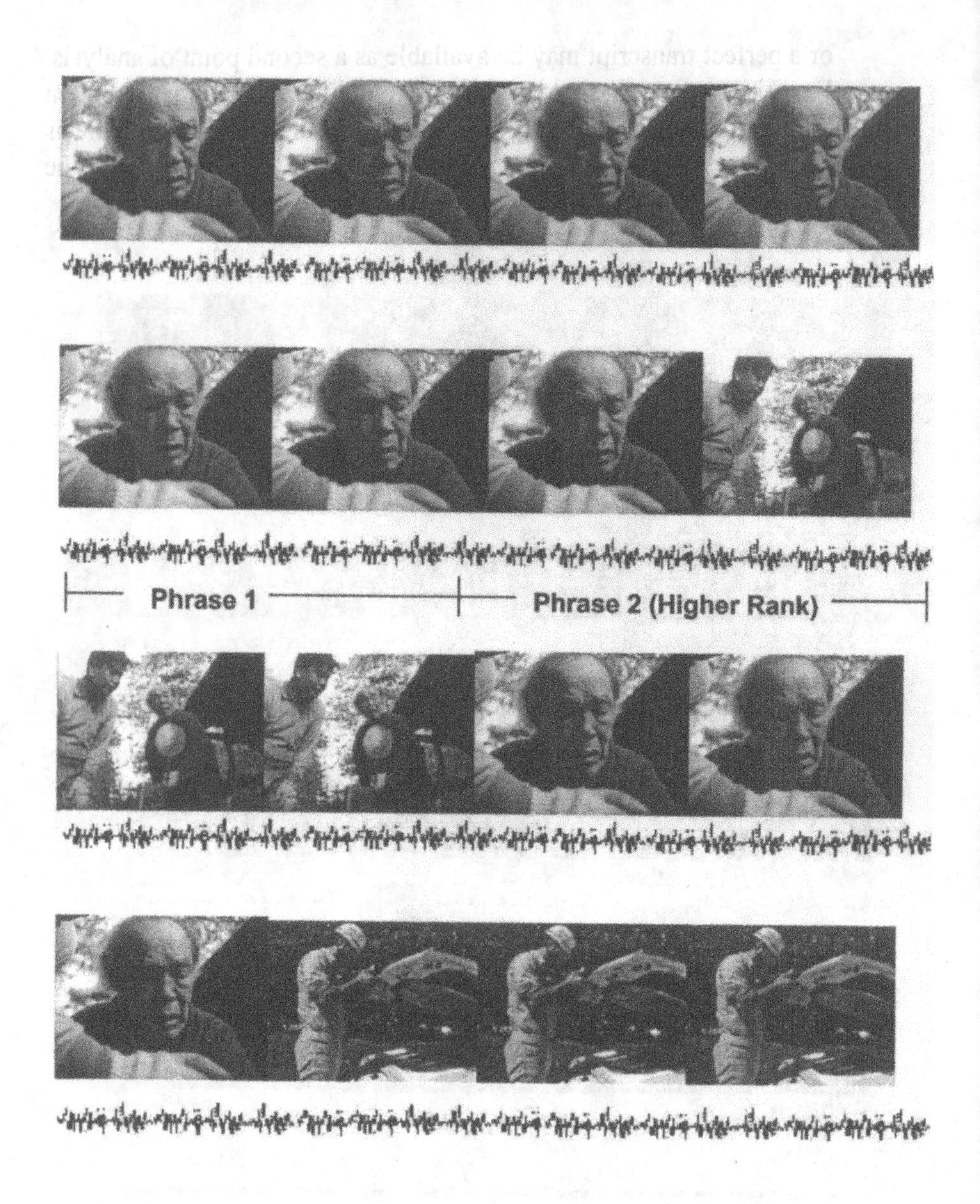

Figure 4.14a Illustration of an Inter-cut shot for image and audio skim selection. The phrase near the Inter-cut has a higher rank than other neighboring phrases.

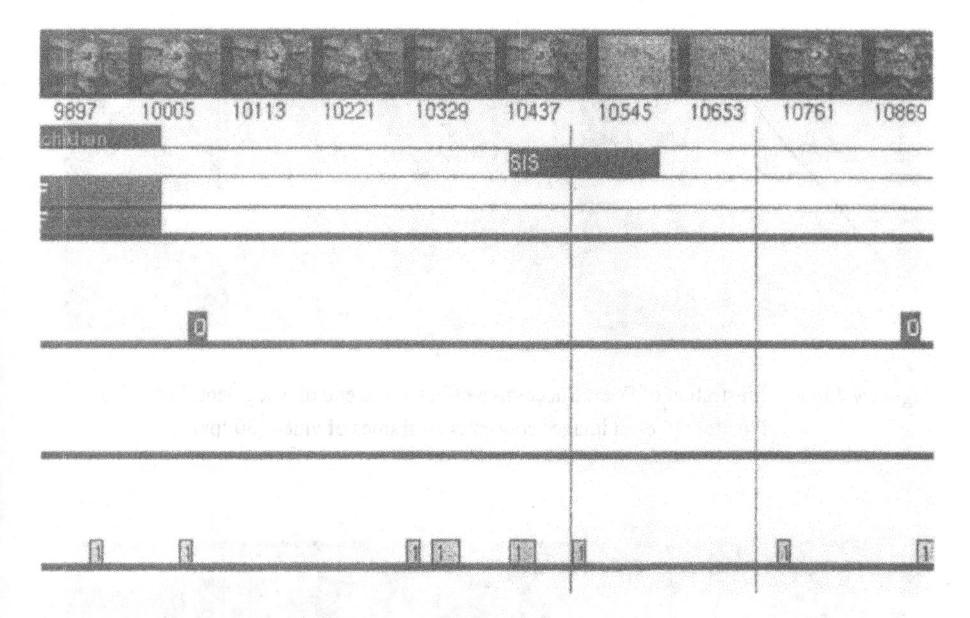

Figure 4.14b Experimental system for detecting Inter-cut shots (SIS).

MR-3 Inter-cut Shots (SIS)

This rule uses an inter-cutting sequence to rank an adjacent audio phrase. The color histogram difference gives us a simple measure for detecting similarity between shots. Shots between successive imagery of a human face usually imply an illustration of the subject. For example, a video producer will often interleave activity shots between shots of a person being interviewed. This effect was illustrated in Figure 2.8. Images that appear between two similar shots, and are less than T_{SS} seconds (typically 8-10 seconds) apart, are characterized as an adjacent similar shot. An illustration of this effect is shown in Figure 4.14a. Audio near the inter-cut receives a higher rank than subsequent audio. Figure 4.14b is the output of a detected inter-cut shot.

MR-4 Short Shots Effect (SS1)

Short successive shots are often used to introduce an important topic. By measuring the duration of each shot, S_D, we can detect these regions and identify short successive sequences. An example of this rule is shown in Figure 4.15a, where each thumbnail lasts approximately 10 frames, or 1/3 a second of video. The first image appears at a normal pace. It is followed by three short shots and the normal pace resumes with the image of a temple. Figure 4.15b is the output of detected short successive shots. This effect was also illustrated in Figure 2.6.

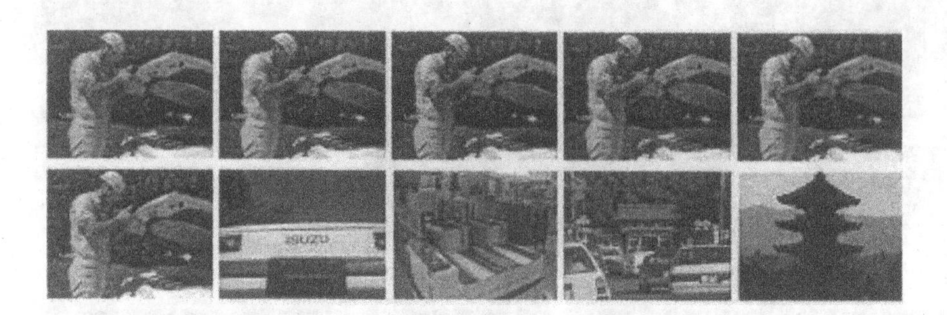

Figure 4.15a Illustration of Short Successive effect at the end of a segment. For illustration, each image represents 10 frames of video (30 fps)

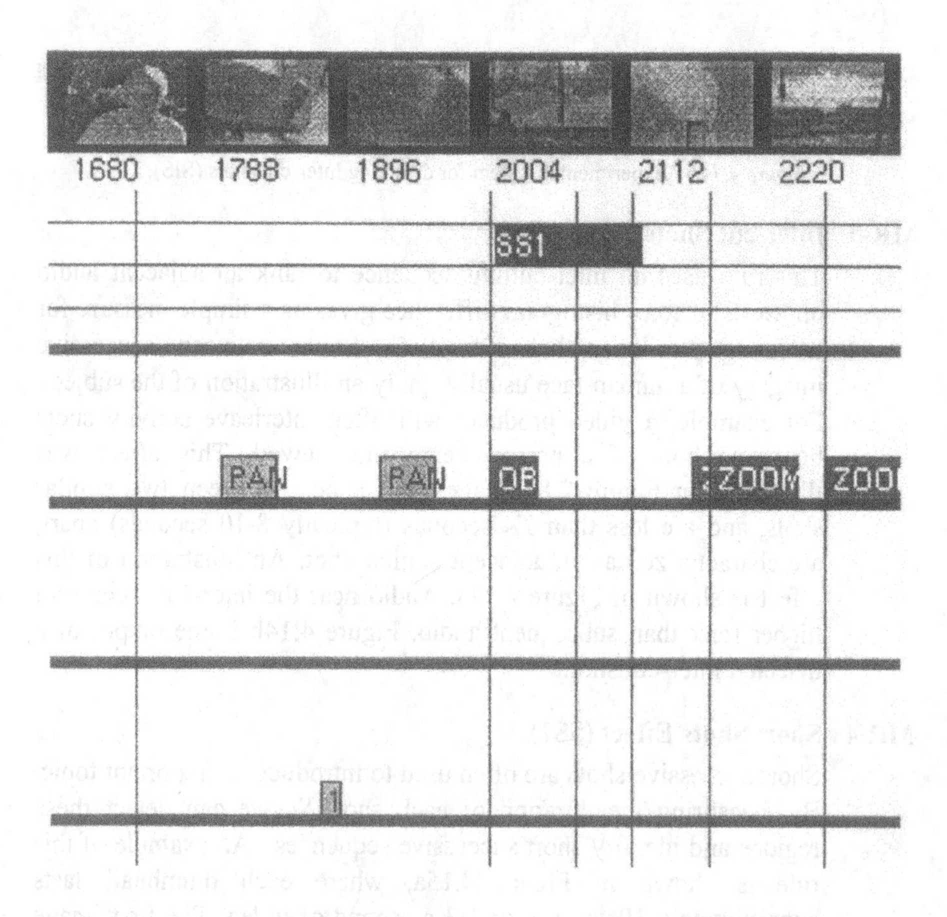

Figure 4.15b Carnegie Mellon Video Skimming system for detecting Short Successive shots (SS1). Keyphrases search for image regions beyond adjacent shots.

4.7 Super-Rules for Genre and Visual Presentation

With prioritized video frames from each shot, we now have a suitable representation for combining the image and audio skims for the final skim. A set of Super-Rules (SR) is used to complete skim creation. The super-rules order these keyphrases and image candidates using the following parameters:

- Video Genre - The type of video (documentaries, broadcast news, feature-films and sports are presented in this chapter).
- Rule Frequency - The frequency at which rules are detected in the video (e.g. more text captions appear in news video than feature-films).
- Compaction - A user specified or preset compaction level that sets skim duration. The duration is not exact due to variations in Keywords and Keyphrases.
- Skim Audio - The audio summary is set for Keyword (1 to 2 seconds) or Keyphrase (5 seconds or more) playback. Shorter skims provide greater compaction, but require Keyword audio. Keyphrase skims are more intelligible, but provide less compaction.
- Duration - The length of the original video (e.g., 5 min news story, 2 hour feature-film).

SR-1 Visual Redundancy

To avoid visual redundancy, we reduce the presence of human faces and default image regions in the skim. If the highest-ranking skim candidate for a keyphrase is the default rule, we extend the search range, W_s, from adjacent shots to one shot back and two shots forward with a temporal search limit of 10 seconds. This will not guarantee improved skim selection, as shown by the image candidates for the keyphrase "reconstruct" in Figure 4.13. The human face rule is used with a smaller number of shots if the segment contains several interviews. Interview scenes can be extremely long, so we extend the search parameters W_s and W_t as before.

SR-2 Frame Rate

The playback frame-rate may be altered to reduce the size of the skim video. The usual frame-rate for MPEG-1 video is 29.97 frames per second. Empirical studies have shown that levels as low as 5 to 6 fps can be achieved without considerable visual degradation

[Christel 1996]. Research in keyframe based browsing representations suggests that frame rates near 10 FPS offer sufficient image comprehension when viewed as video [Ding 1997]. As a minimal frame-rate, we use 15 fps to maintain lip-sync [McGurk 1976]. This is important for news video, which contains multiple anchorperson shots.

SR-3 Multiple Shots

A keyphrase will often extend over several frames of video. When the keyphrase duration, K_d, is longer than 150 frames, images from candidates of adjacent shots are included, with an even distribution of skim frames per shot. This setting can cause a slideshow effect when the duration of each shot is less than 1 or 2 seconds.

Adjacent shots with rapid camera motion should not be used for this rule. The change from a still shot appears disoriented without a subtle transition into the motion shot.

SR-4 Genre Specific Super-Rules

With each type of video, there is a specific order and priority for the primitive and meta-rules. In addition to changes in priority, there are specific rules needed only for a particular type of video. These rules are described for documentaries, broadcast news, sports, feature-films, and general video in the sections that follow.

Genre specific rules describe the degree of even coverage for keyphrases in the skim. Even coverage is appropriate for most video. In broadcast news, however, the coverage should be concentrated at the beginning and end of a segment.

The duration of the audio keyphrase, K_g will also vary between different types of video. The speech in a news broadcast may appear at an accelerated pace compared to a documentary video. The order of primitive rules for each type of video is listed in Table 4.4.

Table 4.3 shows the image selection of different rules for specific types of video. It represents the percentage of shots in a particular genre that correspond to each primitive rule. Certain parameters for skim creation change according to the type of video being processed. Table 4.2 lists the parameters for keyphrase selection and image candidates according to different types of video.

Table 4.2 Parameters for Genre Specific Super-Rules

Video Genre	**Keyphrase Duration Minimum** ***Kg min***	**Keyphrase Duration Maximum** ***Kg max***	**Audio Segment Spacing** ***Sw***	**Keyphrase for Image Shot Splits** ***Kd***	**Image Search Window** ***Wt (Ws)***
Documentary	4 seconds	10 seconds	0.5 seconds	150	5 sec (+ 1 shot)
News	2 seconds	10 seconds	0.2 seconds	120	4 sec (+ 1 shot)
Sports	6 seconds	24 seconds	0.7 seconds	210	2 sec (0 shot)
Feature Films	6 seconds	22 seconds	0.5 seconds	200	4 sec (+ 1 shot)
Default	3 seconds	13 seconds	0.5 seconds	150	5 sec (+ 1 shot)

Table 4.3 Primitive and Meta-Rules Detected for each Video Genre

Video Category	**BCM**	**OBM**	**TXT**	**FAC**	**KNF**	**Intro (Meta -Rule)**	**Other**
Documentary	34.2 %	38.8 %	2.6 %	12.4 %	14.8 %	8.4 %	11.0 %
News	31.4 %	21.8 %	24.3 %	21.9 %	13.7 %	7.8 %	5.2 %
Feature Films	25.6 %	40.2 %	3.0 %	16.5 %	6.4 %	3.1 %	4.5 %
Sports (Football)	28.4 %	18.3 %	6.6 %	8.5 %	2.9 %	1.1 %	7.2 %

SR-4.1 Documentary Rules

As discussed in chapter 2, documentaries usually cover a single topic. The topic is usually introduced at the beginning segment of the video and supported in later segments by related material or historical facts. A summarization segment is included at the end of most documentaries. The keyphrase coverage is relatively even in the middle of a documentary, and slightly smaller at the beginning and the end. A producer will often start a segment by introducing a person of expertise, so the Introduction Rule is important in factual documentaries. The Inter-cut Rule is second in priority, so that video embedded between longer shots is highlighted. Keyphrases near fades or simple cuts are next in priority. In documentaries, fades usually imply a change in content. The remaining rules are less important since they appear with great frequency, such as motion and human faces; or too seldom, such as short sequences and video captions. The documentary rule order is also used as a default setting when the video type is unknown. The rule-base order for the documentary rules is listed in the second column of Table 4.4.

Table 4.4 Rule Order for each Video Genre

Rule Order	Documentaries and Default	Broadcast News	Sports	Feature-Films
1	Introductions	Introductions	Object Motion	Introductions
2	Inter-cuts	Video Text	Camera Motion (Pans)	Keyphrase Near Fade
3	Keyphrase Near Fade	Keyphrase Near Fade	Significant Audio	Bounded Camera Motion
4	Keyphrase Near Cut	Inter-cuts	Keyphrase Near Cut	Significant Audio
5	Short Shots	Short Shots	Human Faces	Video Text
6	Object Motion	Object Motion	Video Text	Short Shots
7	Bounded Camera Motion	Significant Audio	Inter-cuts	Object Motion
8	Video Text	Keyphrase Near Cut	Introductions	Inter-cuts
9	Human Faces	Human Faces	Short Shots	Keyphrase Near Cut
10	Significant Audio	Bounded Camera Motion	Keyphrase Near Fade	Human Faces
11	Default	Default	Default	Default

SR-4.2 Broadcast News Rules

Most producers follow a rigid pattern when editing broadcast news. A segment will usually start with an opening statement by the anchorperson. The beginning and ending portion of a news segment contains a brief summary so the keyphrase coverage is uneven and biased accordingly.

As with documentaries, the introduction of a speaker is considered the most important rule. Video captions are repetitive in news video ticker tape, but they are second in priority when they describe a person, location, or topic. The temporal position of captions will supersede keyphrase position within the search interval, I_w. Similar to documentaries is the importance of a fade or dissolve, which is third in rule priority. The order for the broadcast news rules is listed in column 3 of Table 4.4.

SR-4.3 Sports Rules

Sports video is limited in its visual presentation and order by the rules of the particular type of sport. Each sport may need a different set of rules and potentially a different order of priority. Rules and procedures common to most sports were described in chapter 2. Chapter 2 also describes methods for detecting slow-motion replays as a potential rule for skimming [Kobla 1999].

Object and camera motion imply action, and they are the top two rules for sports. Third is significant audio, where loud sounds from applause or boos may imply an exciting play. The rule order for the sports video is listed in column 4 of Table 4.4.

SR-4.4 Feature-Film Rules

Feature-films vary according to the topic and style of the producer. Rules and procedures common to most feature-films were described in chapter 2. In movies, as with documentaries and broadcast news, introduction scenes top the priority for skim selection. A fade may imply change in content, and is second in priority. Camera motion is used for transitional periods and is important for most films. Loud sounds such as crashes, explosions, and gunfire imply action, and are third in priority for skim selection. Chyron, or intentional video captions almost never appear in films. Subtle texts, such as a clock or street sign, are often shown as a clue to the viewer. This type of captions is the fourth rule in priority. The order for the feature-film rules is listed in column 5 of Table 4.4.

SR-4.5 Unknown Video Genres

The rules for documentaries are the most general, and therefore serve as the default rule order when the type of video is unknown. Commercials, Sitcoms, Dramas, Talk Shows, Interviews, and other types of video ultimately require a custom rule order. The settings for the default case will still provide even coverage and characterization rules for identifying important elements in the video. The default order (Documentary) for unknown video is listed in column 2 of Table 4.4.

4.8 Rule Hierarchy and Tests for Visual Quality

Characterization and summarization rules work in a specific hierarchy. Skim creation starts with the selection of a set of audio keyphrases. Primitive rules then provide image candidates for these keyphrases. Each primitive rule has a priority based on the higher order meta-rules for image selection and video genre. The hierarchy for skim selection is completed with a final stage of super-rules for video clarity. They are the highest order in the selection hierarchy, and they do not integrate image and audio features. The resulting skim must be tested for visual clarity. The tests below improve the visual appearance by removing occlusion and other artifacts from the skim.

Adjustments for Shot Changes

The first test adjusts the visual appearance of the skim according to shot changes. The creation of the skim video involves extracting image regions that may appear close to a shot change. When the start or end of a skim region is too close to a shot change, the overlapping transition will appear short and truncated to the viewer. The minimum keyphrase duration is 60 frames. The 60-frame minimum is based on empirical studies of visual comprehension in short video sequences [Smith 1998].

The final skim borders are adjusted to avoid image regions that overlap or continue into adjacent shots by less than 60 frames. This test for proximity to shot changes uses the start and end portion of a skim region. When the skim end frame overlaps a shot break by less than 60 frames, the image region is shifted forward until the overlap is at least 60 frames. A similar test is used for the start frame of the skim.

Truncation should occur at the beginning of the skim region if necessary. Viewers can better interpolate audio cropped from the beginning of a word than the end. An illustration of the skim boundary adjustment is shown in Figure 4.16. The results of earlier primitive and meta-rules included this adjustment for visual clarity. In Figure 4.13, the DEF image region for the keyword "reconstruct" automatically moved forward to cover the next shot.

Overlapping Image Skims

The final measure for visual clarity tests for overlapping skim image regions. Many of the meta-rules require that image regions be shifted in order to maximize content visualization. These shifts may result in image regions that overlap. When this occurs, the latter region is moved so that the two regions merge for continuous playback. This may cause an overlap into other shots and the check for proximity to adjacent shot changes is always applied after this adjustment.

Skim Testbeds

The results of several automatic and manually created skims are shown in Table 4.5. Manual skims were created for 8 hours of video in the initial stages of these experiments to test for visual clarity and comprehension. The compaction ratio for a typical segment is 10:1, and skims with a compaction ratio as high as 35:1 still retain most of the relevant content. Skims were automatically generated using news and documentary video from the Informedia Digital Video Library [Wactlar 1997] and [Wactlar 1999]. In total, skims were created for over 1,500 hours of original content. These skims were also used in a video library system at the Winchester Thurston K-12 School in Pittsburgh, PA [Christel 1996].

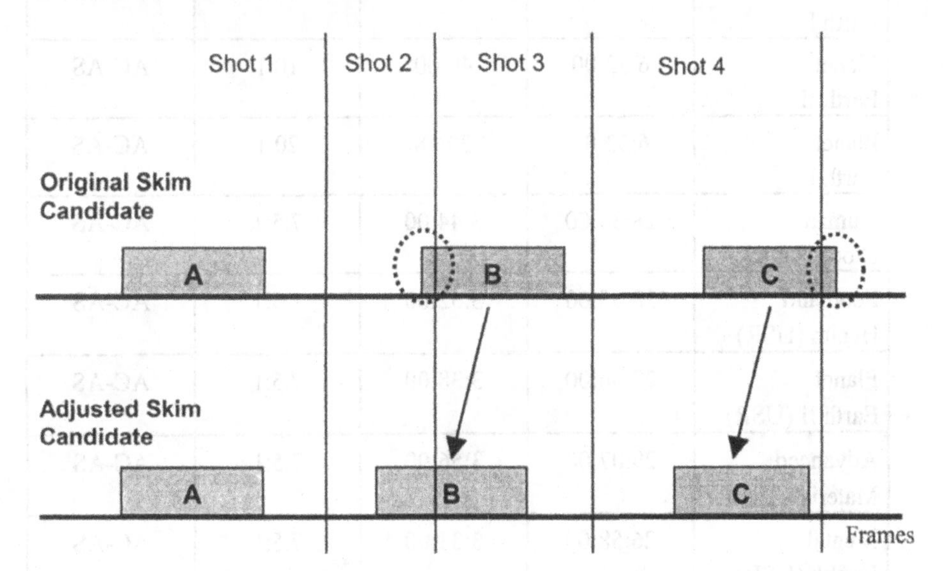

Figure 4.16 Boundaries of image skim candidate regions adjusted to avoid shot changes. (A) This skim region is not close to a shot change – no adjustment necessary (B) Skim region that is too wide to move to shot 3. It is automatically moved to the left to share shots 2 and 3. (C) This skim region is automatically moved to the left since it is small enough to fit in shot 4.

Table 4.5 Video Skimming Compaction Results [Smith 1998]

Title	Original Duration	Skim Duration	Target Compaction	Comments
K'nex, CNN Headline News	61:00	7:13	10:1	MC-AS
Species Destruction I	68:65	6:40	10:1	MC-AS
Species Destruction II	2:03:23	12:43	10:1	MS
International Space Univ.	2:46:20	28:13	6:1	MS
Rain Forest Destruction	1:47:13	5:36	20:1	MS
Mass Extinction	8:19:04	55:05	10:1	AC-AS
Human Archeology	3:31:02	40:08	10:1	AC-AS
Planet Earth I	7:44:05	44:01	10:1	AC-AS
Planet Earth II	6:33:00	40:00	10:1	AC-AS
Planet Earth II	6:33:00	20:18	20:1	AC-AS
Human Clock (USR)	28:34:00	3:44:00	7.5:1	AC-AS
Food and Health (USR)	28:55:00	3:52:00	7.5:1	AC-AS
Planet Earth II (USR)	27:36:00	3:38:00	7.5:1	AC-AS
Advanced Materials (USR)	29:07:00	3:56:00	7.5:1	AC-AS
Mental Health (USR)	26:58:00	3:33:00	7.5:1	AC-AS

Comments

MC - Manually Assisted Characterization
AC - Automated Characterization
MS - Manual Skim Creation
AS - Automated Skim Creation

Experimental Results

The previous sections showed experimental results for the rules associated with skim creation. The system for these experiments use the methods in chapters 2 and 3 to automatically create video skims at user specified compaction levels. An example of a full skim is shown in Figure 4.17. The images are snapshots of the video skim at intervals of approximately 60 frames. There is limited image redundancy because the system has split long shots to include imagery from multiple shots. Note the intelligibility of the audio, or transcript portion of the skim. The phrases are cropped, but concise in grammar.

It is important to compare the results of rule-based skims with other forms of skim creation. The first comparison is based on simplistic methods for extracting important audio and images. In Figure 4.18, a skim is shown with audio and image regions that are extracted at fixed intervals. Like Figure 4.17, the images are snapshots of the video skim at intervals of approximately 60 frames. There is visual redundancy because an image selection based on fixed intervals does not split long shots into multiple shots. Note the unintelligibility of the audio, or transcript portion of the skim. The phrases are cropped as before, but not concise in grammar. In some cases, only a single word is selected for a given interval. The playback of this type of skim is usually very difficult to comprehend. The results of several experimental skim formats and user-studies are discussed in chapters 5 and 6.

4.9 Conclusions

This chapter describes a rule-base system for video characterization and summarization applications. Sections 4.1 and 4.2 define video skims and other useful video surrogates. Section 4.3 and 4.4 describe the selection of significant audio and images using the features from chapter 3. Primitive rules are derived in section 4.5 from experimentation with a variety of different video genres and production standards. Meta-rules and Super-rules that integrate multiple features for video summarization were developed in sections 4.6 and 4.7. Methods for testing visual skim quality and setting rule hierarchies were described in section 4.8.

Chapter 5 extends the methods in video summarization to other formats and interfaces. Chapter 6 evaluates these summaries and interfaces through a series of user-studies.

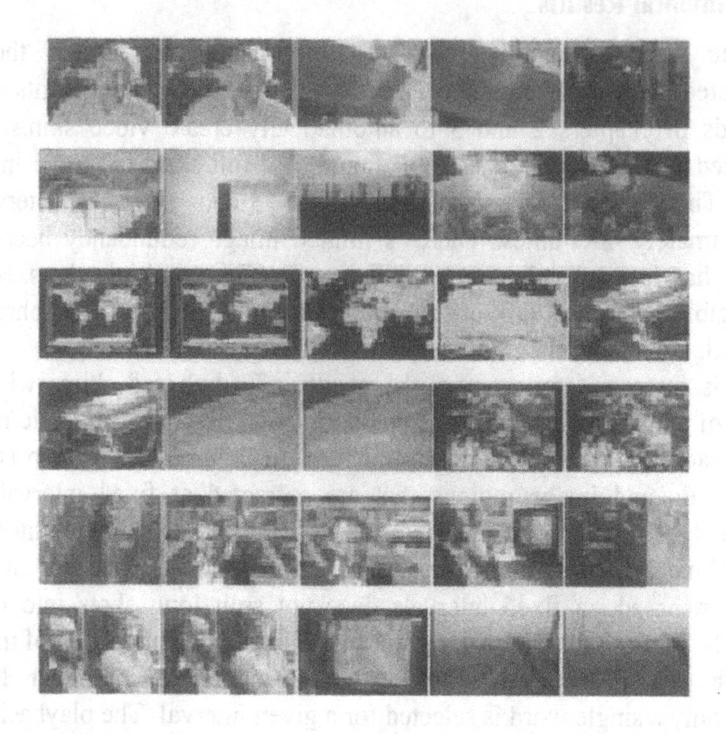

Transcript:

Our research show us that mass extinctions are relatively common | this mass extinction is not triggered by some extraterrestrial phenomena | man's destruction of the diversity of life | man has the technology to change the world | human waste are threatening the web of life | a tapestry of lights track the human presence | fires in africa fuel the struggle against famine | at nasa's goddard space flight center in maryland (pause) tucker draws | past eight years tucker has observed subtle shifts | discovery returns to space on a mission to photograph the earth | you see these incredibly large fires | compton tucker has been monitoring these fires | the destruction was triggered | homesteaders are transforming the wilderness | norman myers has voiced his concern | invited myers to review his most recent findings | we could have 2 or 3 smaller fires burning | we're pushing species down the tubes on our own planet | what splendid creatures once lived here

Figure 4.17 Skim video frames and keyphrases for "Planet Earth - I" (10:1 compaction). Most scenes are represented without redundant imagery. All audio phrases are intelligible and contain at least four audible words.

Transcript:

our research show us that mass extinctions are relatively
and diversity lost
and in fact may be regarded as a human meteorite
stripped away for resources
against famine in the third world | at nasa's goddard space flight
houston discovery | amazon they are stunned by what they see
image confirm that in one | roads | visible from space | as
is particularly disturbing | fires there at about 2:45 in the afternoon
more probes sent to off to have a look at this little
in the long run... it matters enormously
a few becomes many becomes too many

Figure 4.18 Fixed interval selection video frames and keyphrases for "Planet Earth - I" (10:1 compaction). Many scenes appear at great length, while others have little content. The Audio portion contains many unintelligible phrases, and in some cases, single or no words at all.

4.10 Bibliography

Boreczky, J.S., and Rowe, L. A. "Comparison of Video Shot Boundary Detection Techniques", *Storage and Retrieval for Still Image and Video Databases IV*, Proc. SPIE 2664, pp. 170-179, Jan. 1996.

Christel M.G., and Pendyala K. "Informedia Goes to School: Early Findings from the Digital Video Library Project". *D-Lib Magazine*, September 1996, www.dlib.org/dlib/september96/informedia/09christel.html.

Christel, M., Stevens, S., Kanade, T., Mauldin, M., Reddy, R., & Wactlar, H. "Techniques for the Creation and Exploration of Digital Video Libraries". Chapter 8 of Multimedia Tools and Applications, B. Furht, ed. Kluwer Academic Publishers, 1996.

Christel, M.G., Smith, M.A., Taylor, C.R, and Winkler D.B., "Evolving Video Skims into Useful Multimedia Abstractions". *Proceedings of the CHI '98 Conference on Human Factors in Computing Systems*, C. Karat, A. Lund, J. Coutaz and J. Karat, Eds. (Los Angeles, CA, April 1998), pp. 171 - 178.

Ding, W., Marchionini, G., and Tse, T., "Previewing Video Data: Browsing Key Frames at High Rates Using a Video Slide Show Interface." *Proceedings of the International Symposium on Research, Development and Practice in Digital Libraries,* Tsukuba Science City, Japan, November 1997.

Firmin T., and Chrzanowski M.J., "An Evaluation of Automatic Text Summarization Systems", Maybury, M.T. editor, Advances in Automatic Text Summarization, The MIT Press, Cambridge, MA, 1999.

Forsyth, D.A., Haddon, J., and S. Ioffe, "Finding objects by grouping primitives," in *Shape, contour and grouping in computer vision* , D.A. Forsyth, J. Mundy, R. Cipolla and V. DiGes'u, Eds., Springer-Verlag LNCS 1681, 2000.

Furht, B., editor, "Multimedia Tools and Applications", Kluwer Academic Publisher, Norwell, MA, 1996.

Furht, B., Editor-in-Chief, "Handbook of Multimedia Computing", CRC Press, Boca Raton, Florida, 1999.

Gargi, U., Kasturi, R., Strayer. S.H., "Performance Characterization of Video-Shot-Change Detection Methods". *IEEE Transactions on Circuits and Systems for Video Technology*, Vol. 10, No. 1, February 2000.

Hampapur, A., Jain, R., and Weymouth, T. Production Model Based Digital Video Segmentation. *Multimedia Tools and Applications*, 1 (March 1995), 9-46.

Hauptmann, A.G., Speech Recognition in the Informedia Digital Video Library: Uses and Limitations, ICTAI-95 7th *IEEE International Conference on Tools with AI*, Washington, DC, November 6-8, 1995.

Jarina R, Murphy N, O'Connor N and Marlow S. "Speech-Music Discrimination from MPEG-1 Bitstream", *SSIP'01 - WSES International Conference on Speech, Signal and Image Processing*. Malta, 1-6 September 2001.

Kedem, B., "Spectral Analysis and Discrimination by Zero-Crossings", *IEEE, Proceedings, vol 74 no 11, November 1986.*

Kobla, V., DeMenthon, D., Doermann, D., "Detection of slow-motion replay sequences for identifying sports videos", *Multimedia Signal Processing*, 1999 IEEE 3rd Workshop, Page(s): 135 –140.

Li, F., Gupta, A., E., Sanocki, L., He, L., and Rui, Y., "Browsing Digital Video", in *Proceedings of ACM CHI* (The Hague, Netherlands, April 2000), ACM Press.

Lienhart, R., "Reliable Transition Detection In Videos: A Survey and Practitioner's Guide". MRL Technical Report; *International Journal of Image and Graphics* (IJIG), August 2001.

McGurk, H., and MacDonald, J., "Hearing lips and seeing voices," *Nature*, pp. 746-748, December 1976.

"MPEG Digital Video Standard", Motion Picture Experts Group 2003, http://mpeg.telecomitalialab.com/.

Saunders,J., Real-Time "Discrimination of Broadcast Speech/Music", *ICASSP 96, vol. 2 (pp 993-996).*

Schneiderman, H., and Kanade, T., "Object Detection Using the Statistics of Parts ", *International Journal of Computer Vision*, 2002.

Sleator, D., and Temperley, D., "Parsing English with a Link Grammar," *Third International Workshop on Parsing Technologies,* 1993.

Smith, M.A., and Kanade, T., "Video Skimming for Quick Browsing Based on Audio and Image Characterization." Carnegie Mellon University Technical Report CMU-CS-95-186, July 1995.

Smith, M., Kanade, T. "Video Skimming and Characterization through the Combination of Image and Language Understanding Techniques". *Computer Vision and Pattern Recognition*, San Juan, Puerto Rico, June 17-19, 1997.

Smith. M., "Integration of Image, Audio, and Language Technology for Video Characterization and Variable-Rate Skimming", PhD Thesis, Department of Electrical and Computer Engineering, Carnegie Mellon University. January 1998.

Wactlar, H., Hauptmann, A., Smith, M.A., Pendyala, K., Garlington, D. "Automated Video Indexing of Very Large Video Databases," *SMPTE Journal*, August, 1997.

Wactlar, H., Christel, M., Gong, Y., and Hauptmann, A. "Lessons Learned from the Creation and Deployment of a Terabyte Digital Video Library". *IEEE Computer* 32(2), February 1999, 66-73.

Chapter 5
Visualization Techniques

Co-authored by Michael G. Christel
Department of Computer Science, Carnegie Mellon University

Visualization techniques provide user interfaces for applications in video characterization. They describe content and present video for specific types of summarization applications. Video skims, poster frames, thumbnails, text titles, and segmentation applications are described in this chapter. Controls for summarization parameters provide customization in video interfaces. The Informedia Digital Video Library at Carnegie Mellon University is used as a platform for presenting the various visualization techniques in this chapter.

5.1 Visualization Categories

The format for video summary presentation is not limited to shortened viewing of a single story. Several alterations to the audio and image selection parameters may result in drastically different summaries during playback. Summary presentations are usually visual and textural in layout. Textual presentations provide more specific information and are useful when presenting large collections of data. Visual, or iconic, presentations are more useful when the content of interest is easily recalled from imagery. This is the case in stock footage video where there is no audio to describe the content. Many of the common formats for summarizing video are described below.

Titles – Text Abstracts

Text titles have a long history of research through text abstraction and language understanding [Mauldin 1989]. The basic premise is to represent a document with a single sentence or phrase. A transcript, closed-captions, production notes, or on-screen text provide the text from which one can select a title in video.

Thumbnails and Storyboards - Static Filmstrips

The characterization analysis described in chapter 3 for selecting important image and audio in summaries may be applied to select static poster-frames. These frames may be shown as a single representative image, or thumbnail, or as a sequence of images over time, such as a storyboard.

Variable Duration Summaries

For specific applications, a compaction level is used during playback to control the duration of the video summary. This feature is similar to the speed control roll bar in videocassette editing systems. It provides variable rate summarization and duration control as a parameter in the user interface.

Skims – Dynamic Summaries

The automated skim was defined in Chapter 4 [Smith 1998]. Skims provide summarization in the form of moving imagery. They are subsets of an original sequence placed together to form a shorter video. The video skim does not always contain an audio track.

User Controlled Summaries

For video libraries and indexing applications, an interactive weighting system uses text queries to customize video summaries. A text query is entered and used for the TF-IDF weighting process in keyword extraction.

Words from the user's query receive a higher weight and are automatically selected for the video summary. This provides a summary based on the rules described in chapter 4 and user's interests.

Summarizing Multiple Documents

A summary may cover more than a single document or video. It can encompass several sets of data at an instant or over time. In the case of news, it is necessary to analyze multiple stories to accurately summarize the events of a day.

5.2 Evaluation Methods for Visualization

The Informedia Project at Carnegie Mellon University has created a multi-terabyte digital video library consisting of thousands of hours of video, segmented into tens of thousands of documents. Since Informedia's inception in 1994, numerous interfaces have been developed and tested for accessing this library, including work on multimedia surrogates that represent a video document in an abbreviated manner [Christel 1999] and [Wactlar 1999]. The video surrogates build from automatically derived descriptive data, i.e., metadata, such as transcripts and representative thumbnail images derived from speech recognition, image processing, and language processing. Through human computer interaction evaluation techniques and formal user studies, the surrogates are tested for their utility as indicative and informative summaries, and iteratively refined to better serve user's needs. This section discusses evaluations for text titles, thumbnail images, storyboards, and skims.

Various evaluation methods are employed with the video surrogates, ranging from interview and freeform text user feedback to "discount" usability techniques on prototypes to formal empirical studies conducted with a completed system. Specifically, transaction logs are gathered and analyzed to determine patterns of use. Text messaging and interview feedback allow users to comment directly on their experiences and provide additional anecdotal comments. Formal studies allow facets of surrogate interfaces to be compared for statistically significant differences in dependent test measures such as success rate and time on task.

Discount usability techniques, including heuristic evaluation, cognitive walkthrough, and think-aloud protocol, allow for quick evaluation and refinement of prototypes [Nielson 1994]. Heuristic evaluation lets usability specialists review an interface and categorize and justify problems based on established usability principles, i.e., heuristics. With cognitive walkthrough, the specialist takes a task, simulates a user's problem-solving process at each step through the interaction, and checks if the simulated user's goals and memory content can be assumed to lead to the next correct action. With think-aloud protocol, a user's interaction with the system to accomplish a given task is videotaped and analyzed, with the user instructed to "think aloud" while pursuing the task.

5.3 Text Titles

A text title can be associated with most news and documentary videos. The spoken narrative can be deciphered with speech recognition and the resulting text is time-aligned to the video [Wactlar 1999]. Additional text can also be generated through "video OCR" processing, which detects and translates the text overlaid on video frames into ASCII format [Sato 1998]. From this set of words for a video, a text title can be automatically extracted, allowing for interfaces as shown in Figure 5.1, where the title for the third search result video is shown. Given a query for "Collin Powell trip", the interface shows the search result set in the order of highest word occurrence from the videos. These text titles act as informative summaries, a label for the video that remains constant across all query and browsing contexts. Hence, a query on OPEC that matches the third result shown in Figure 5.1 would present the same title.

The most significant words, as determined by the highest TF-IDF value, can be extracted from a video's text metadata and used as the title. Feedback from an initial user group of teachers and students at a nearby high school in Pittsburgh, PA, showed that the text title was referred to often, and used as a label in multimedia essays and reports, but that its readability needed to be improved. This feedback took the form of anecdotal reporting through email and a commenting mechanism located within the library interface, shown as the pull-down menu "Comments!" in Figure 5.1. Students and teachers were also interviewed for their suggestions, and timed transaction logs were generated of all activity with the system, including the queries, which surrogates were viewed and what videos were watched [Christel 1996].

The title surrogate was improved by extracting phrases with high TF-IDF score, rather than individual words. Such a title is shown in Figure 5.1. The title starts with the highest scoring TF-IDF phrase. As space permits, the highest remaining TF-IDF phrase is added to the list, and when complete the phrases are ordered by their associated video time (e.g., dialogue phrases are ordered according to when they were spoken). In addition, user feedback noted the importance of reporting the copyright/production date of the material in the title, and the size in standard hours:minutes:seconds format. The modified titles were well-received by the users, and phrase-based titles remain with the Informedia video library today, improved by new work in statistical analysis and named entity extraction.

Figure 5.1 shows how the text title provides a quick summary in readable form. Being automatically generated, it has flaws. For example, it would be better to resolve pronoun references such as "he" and it would do better with appropriate upper and lower case. The flaws with any particular surrogate, however, can be compensated with a number of other surrogates or alternate views into the video. For example, in addition to the informative title it

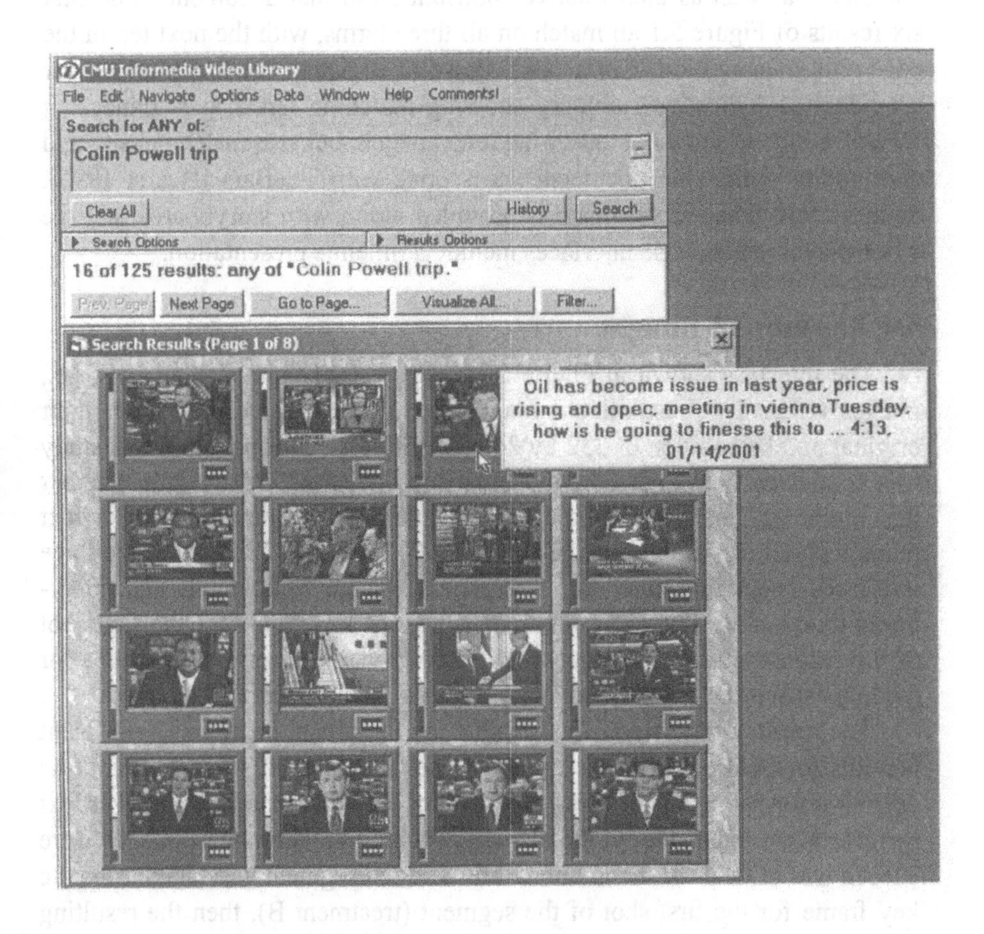

Figure 5.1 Informedia User Interface. Given a query at the top ("Collin Powell trip"), it shows the search results as thumbnail surrogates in the order of word occurrence. Videos containing a larger number of query words appear higher in the search results. A text title can be viewed to show additional detail, as shown in the third search result video.

would be nice to have an indicative summary showing which terms match for any particular video document.

With Figure 5.1, the results are ordered by relevance as determined by the query engine. For each result, there is a vertical "relevance thermometer" bar that is filled in according to the relevance score for that result. The bar is filled in with colors that match with the colors used for the query terms. In this case, with "Colin" colored red, "Powell" colored violet, and "trip" colored blue, the bar shows at a glance which terms match a particular document, as well as their relative contribution to that document. The first six results of Figure 5.1 all match on all three terms, with the next ten in the page only matching on "Colin" and "Powell." Hence, the thermometer bar provides the indicative summary showing matching terms at a glance. It would be nice to show not only what terms match, but also match density and distribution within the document, as is done with TileBars [Hearst 1995]. Such a more detailed surrogate is provided along with storyboards and the video player, since those interfaces include a timeline presentation.

5.4 Thumbnail Images

The interface shown in Figure 5.1 makes use of thumbnail images, i.e., images reduced in resolution by a quarter in each dimension from their original pixel resolution of 352 by 240 (MPEG-1). A formal empirical study was conducted with high school scholarship students to investigate whether this thumbnail interface offered any advantages over simply using a text menu with all of the text titles [Christel 1997]. Figure 5.2 illustrates the three interfaces under investigation: text menus, "naïve" thumbnails and query-based thumbnails. The naïve thumbnails use the key frame for the first shot of the video document, and query-based thumbnails select the key frame for the highest scoring shot for the query, as described earlier in Chapter 4.

The results of the study showed that the thumbnail menu had significant benefits for both performance time and user satisfaction: subjects found the desired information in 36% less time with certain thumbnail menus over text menus. Most interestingly, the manner in which the thumbnail images were chosen was critical. If the thumbnail for a video segment was taken to be the key frame for the first shot of the segment (treatment B), then the resulting pictorial menu of "key frames from first shots in segments" produced no benefit compared to the text menu. Only when the thumbnail was chosen based on usage context, i.e., treatment C's query-based thumbnails, was there an improvement. When the thumbnail was chosen based on the query, by using the key frame for the shot producing the most matches for the query, then pictorial menus produced clear advantages over text-only menus [Christel 1997].

Search Results

[1.00] INV0641 The Great Dinosaur Hunt in Paris, Gould consid
[0.47] INV0638 Nanotyrannus resembles Troodon, and many livi
[0.30] INV071 Environmental destruction causes extinction of
[0.30] INV061 The study of dinosaurs has been dynamic over
[0.24] INV069 Paleontologists recreate and display the creatu
[0.21] INV0637 The nanotyrannus was similar to the Tyrannosa
[0.20] INV0629 Today scavengers at flood sites are birds, maki
[0.20] INV066 In the 18th century, extinction was a controver
[0.17] INV0636 One can find new species and ideas on the she
[0.17] PLE0436 A new theory links the extinction of dinosaurs t
[0.16] PLE0444 The existence of Nemesis may illuminate human
[0.14] INV0642 Dinosaurs roamed the earth for far longer than r

A. Text Menu

Search Results

B. "Naïve" Thumbnails

C. Query-based Thumbnails

Three summary treatments used in empirical study, each treatment always representing 12 video documents

(Note: views are scaled down here to fit into a single figure)

Figure 5.2 Snapshots of the three summary treatments used in the study for evaluating the benefit of thumbnails with 30 subjects.

This empirical study validated the use of thumbnails for representing video segments in query result sets. It also provided evidence that leveraging from multiple processing techniques leads to improvements in the digital video library interface. Thumbnails derived from image processing alone, such as choosing the image for the first shot in a segment, produced no improvements over the text menu. However, the combined use of speech recognition, natural language processing and image processing results in improvements. Via speech recognition the spoken dialogue words are tightly aligned to the video imagery. Through natural language processing the query is compared to the spoken dialogue, and matching words are identified in the transcript and scored. Via the word alignment, each shot can be scored for a query, and the thumbnail can then be the key frame from the highest scoring shot for a query. The result sets showing such query-based thumbnails produce advantages over text-only result presentations and serve as useful indicative summaries.

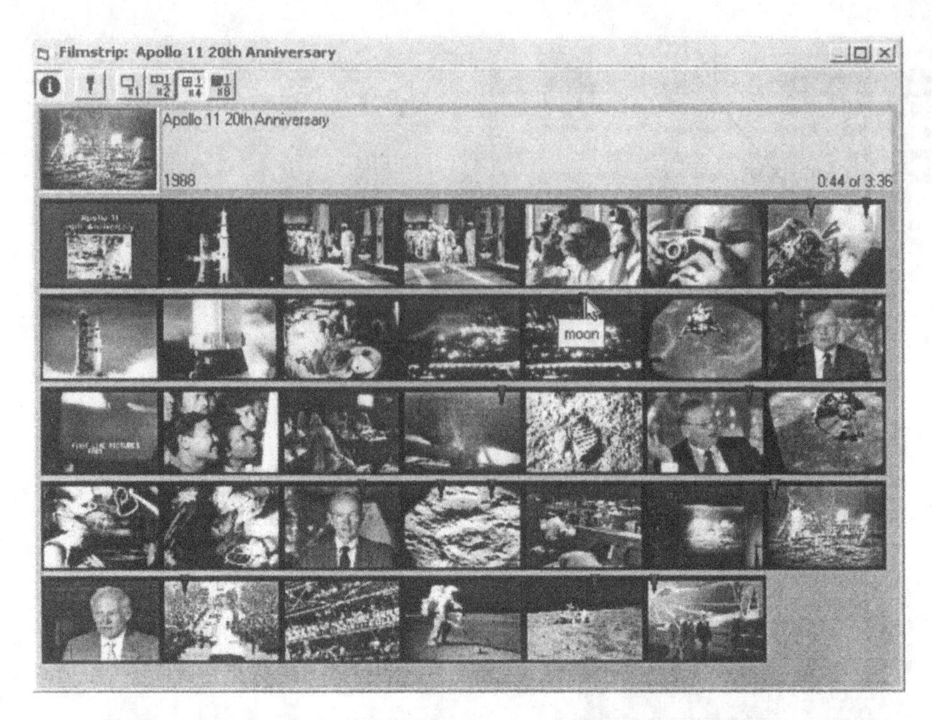

Figure 5.3 Storyboard, with overlaid match "notches" following query on "man walking on the moon."

5.5 Storyboards

Rather than showing only a single image to summarize a video, another common approach presents an ordered set of representative thumbnails simultaneously on a computer screen [Hampapur 1995], [Komlodi 1998], [Taniguchi 1997], [Yeo 1997], and [Zhang 1995]. An example of this storyboard interface, also referred to in the Informedia library as "filmstrip", is shown for a video clip in Figure 5.3. Storyboards address certain deficiencies of the text title and single image surrogates. Text titles and single thumbnail images (Figure 5.1) serve as quick communicators of video segment content, but do not present any temporal details. The storyboard surrogate communicates information about every shot in a video segment. Each shot is represented in the storyboard by a single image, or key frame. Within the Informedia library interface, the shot's middle frame is assigned by default to be the key frame. If camera motion is detected and that motion stops within the shot, then the frame where the camera motion ends is selected instead. Other image processing techniques, such as those that detect and avoid low-intensity images and those favoring images where faces or overlaid text appears, further refine the selection process for key frames [Schneiderman 2002].

Figure 5.4 Reduced storyboard display for same video represented by full storyboard in Figure 5.3.

As with the thermometer bar of Figure 5.1, the storyboard indicates which terms match by drawing color-coded match "notches" at the top of shots. The locations of the notches indicate where the matches occur, showing match density and distribution within the video. Should the user mouse over a notch, as is done in the twelfth shot of Figure 5.3, the matching text for that notch is shown, in this case "moon." As the user mouses over the storyboard, the time corresponding to that storyboard location is shown in the information bar at the top of the storyboard window, e.g., the time corresponding to the mouse location inside of the twelfth shot in Figure 5.3 is 44 seconds into the 3 minute and 36 second video clip. The storyboard summary facilitates quick visual navigation. To jump to a location in the video, e.g., to the mention of "moon" in the twelfth shot, the mouse can be clicked at that point in the storyboard.

One major difficulty with storyboards is that there are often too many shots to display in a single screen. There are attempts to reduce the number of shots represented in a storyboard to decrease screen space requirements [Boreczky 2000], [Lienhart 1997], and [Yeo 1997]. In Video Manga [Uchihashi 1999], and [Boreczky 2000], the interface presents thumbnails of varying resolutions, with more screen space given to the shots of greater importance.

In the Informedia storyboard interface, the query-based approach is again used: the user's query context can indicate which shots to emphasize in an abbreviated display. Consider the same video document represented by the storyboard in Figure 5.3. By only showing shots containing query match terms, only 11 of the 34 shots need to be kept. By reducing the resolution of each shot image, screen space is further reduced. In order to see visual detail, the top information bar can show the storyboard image currently under the mouse pointer in greater resolution, as illustrated by Figure 5.4. In this figure, the mouse is over the ninth shot at 3:15 into the video, with the storyboard showing only matching shots at 1/8 resolution in each dimension.

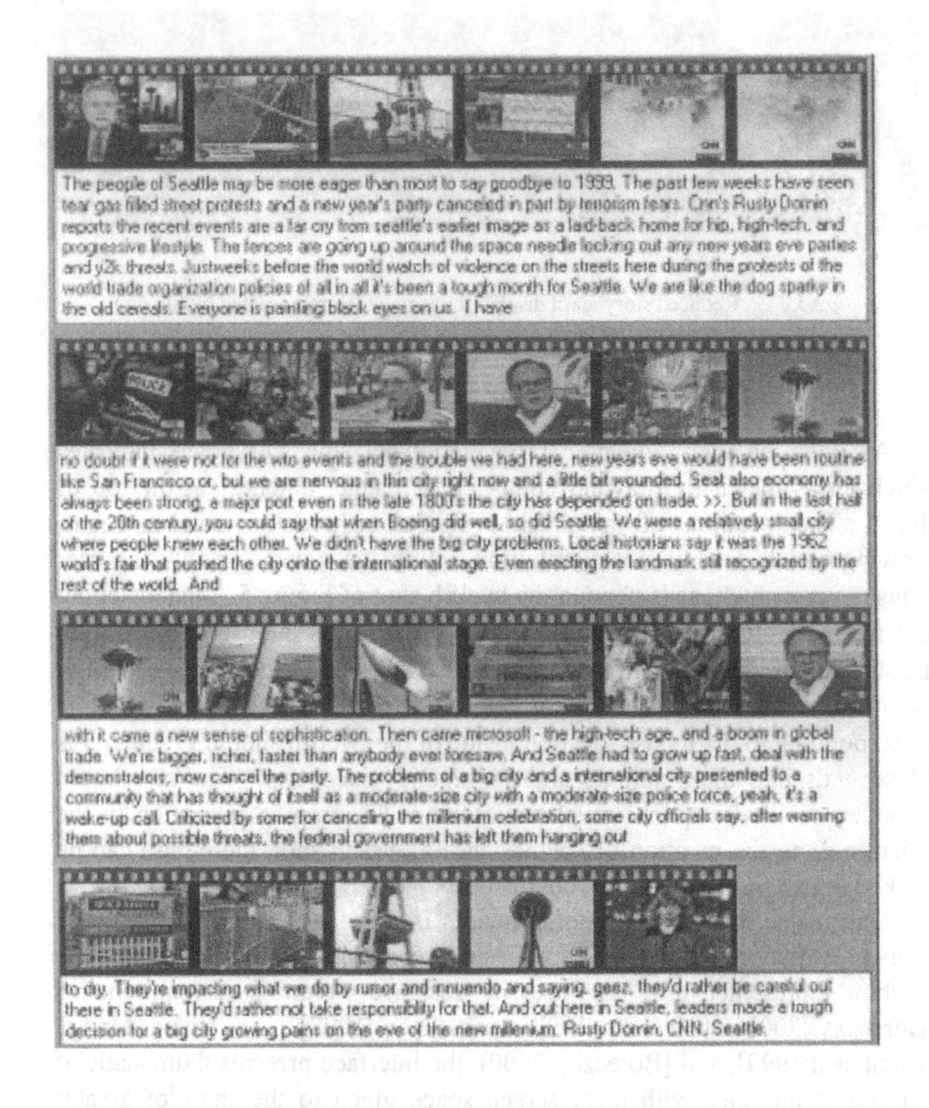

Figure 5.5 **Scaled-down view of storyboard with full transcript text aligned by image row.**

5.6 Storyboard Plus Text

Another problem with storyboards is that their effectiveness varies depending on the video genre [Li 2000]. For visually rich genres like travelogues and documentaries, they are very useful. For classroom lecture or conference presentations, however, they are far less important. For a genre such as news video, in which the information is conveyed both by visuals (especially field footage) and audio (such as the script read by the newscaster), a mixed presentation of both synchronized shot images and transcript text may offer benefits over image-only storyboards. Such a "storyboard plus text" surrogate is shown in Figure 5.5.

This "storyboard plus text" surrogate led to a number of questions:

- Does text improve navigation utility of storyboards?
- Is less text better than complete dialogue transcripts?
- Is interleaved text better than block text?

For example, the same video represented in Figure 5.5 could have a concise storyboard-plus-text interface in which the text corresponding to the video for an image row is collapsed to at most one line of phrases, as shown in Figure 5.6.

To address these questions, an empirical user study was conducted in May 2000 with 25 college students and staff [Christel 2001]. Five interface treatments were used for the experiment: image-only storyboard (like Figure 5.3), interleaved full text (Figure 5.5), interleaved condensed text (Figure 5.6), block full text (in which image rows are all on top, with text in a single block after the imagery), and block condensed text.

Figure 5.6 Storyboard plus concise text surrogate for same video clip represented by Figure 5.5.

The results from the experiment showed with statistical significance that text clearly improved utility of storyboards for a known item search into news video [Christel 2001]. This was in agreement with prior studies showing that the presentation of captions with pictures can significantly improve both recall and comprehension, compared to presenting either pictures or captions alone [Nugent 1983], [Hegarty 1993], and [Large 1995]. Interleaving text with imagery was not always best: even though interleaving the full text by row like Figure 5.5 received the highest satisfaction measures from subjects, their performance for finding known items was relatively low.

The shortest navigation times were accomplished with the interleaved condensed text and the block full text. Hence, reducing text is not always the best surrogate for faster task accomplishment. However, given that subjects preferred the interleaved versions and that condensed text takes less display space than blocking the full transcript text beneath the imagery, the experiment concluded that an ideal storyboard plus text interface is to condense the text by phrases, and interleave such condensed text with the imagery.

5.7 Skim Visualizations

While storyboard surrogates represent the temporal dimension of video, they do so in a static way: transitions and pace may not be captured, and audio cues are ignored. The idea behind a "video skim" is to capture the essence of a video document in a collapsed snippet of video, e.g., representing a 5 minute video as a 30 second video skim that serves as an informative summary for that longer video. Skims are highly dependent on genre: a skim of a sporting event might include only scoring or crowd-cheering snippets, while a skim of a surveillance video might include only snippets where something new enters the view.

Skims of educational documentaries were studied in detail by Informedia researchers as the digital library interface. Users accessed skims as a comprehension aid to understand quickly what a video was about. They did not use skims for navigation, e.g., to jump to the first point in a video where a particular topic is discussed. Storyboards serve as much better navigation aids because there is no temporal investment that needs to be made by the user; for skims, the user must play and watch the skim.

For documentaries, the audio narrative contains a great deal of useful information. Early attempts at skims did not preserve this information well. Snippets of audio for an important word or two were extracted and stitched together in a skim. This was received poorly by users, much like early text titles comprised of the highest TF-IDF words were rejected in favor of more readable concatenated phrases. By extracting audio snippets marked by silence boundaries, the audio portion of the skim became greatly improved, as the skim audio became more comprehensible and less choppy.

A user study was conducted to investigate the importance of aligning the audio with visuals from the same area of the video. Specifically, it was believed that skims composed of larger snippets of dialogue would work better than shorter snippets, the equivalent of choosing phrases over words. A new skim was developed that comprised of snippets of audio bounded by significant silences, more specifically audio signal power segmentation [Christel 1998]. The transcript text for the audio snippets was ranked by TF-IDF values and the highest valued audio snippets were included in the skim, with the visual portion for the skim snippets being in the close neighborhood of the audio.

Another user study was conducted to compare the utility of different types of skims as informative summaries of video. Several students have tested various treatments of skims. Following a playing of either a skim or the full video, the subject was asked which of a series of images were seen in the video just played, and which of a series of text summaries would make sense as representing the full source video. As expected, the FULL treatment performed best, i.e., watching the full video is an ideal way to determine the information content of that full video. The subjects preferred the full video to any of the skim types. However, subjects favored the best automated skim over the subsampled skim treatments, as indicated by subjective ratings. These results are encouraging; incorporating speech, language, and image processing into skim video creation produces skims that are more satisfactory to users. The results and method of evaluation for this study are described in chapter 6.

5.8 Lessons from Single Document Surrogates

In summary, usage data, HCI techniques and formal experiments have led to the refinement of single document video surrogates in the Informedia digital video library over the years. Thumbnail images are useful surrogates for video, especially as indicative summaries chosen based on query-based context. The image selection for thumbnails and storyboards can be improved via camera motion data and corpus-specific rules. For example, in the news genre shots of the anchorperson in the studio and weather reporter in front of a map typically contribute little to the visual understanding of the news story. Such shots can be de-emphasized or eliminated completely from consideration as single image surrogates in storyboards.

Text is an important component of video surrogates. Text titles of a video document serve as a quick identifier. Adding synchronized text to storyboards helps if interlaced with the imagery. Assembling from phrases (longer chunks) works better than assembling from words (shorter chunks).

Showing distribution and density of match terms is useful, and can naturally be added to a storyboard or the progress bar of a video player. The interface representation for the match term can be used to navigate quickly to that point in the video where the match occurs.

Finally, in a temporal summary like skims, transition points between extracted snippets are important. Skims with audio breaks at silence points are received much better than skims with abrupt, choppy transitions between audio snippets.

As digital video assets grow, so do result sets from queries against those video collections. In a library of a few thousand hours of video comprising tens of thousands of video documents, many queries return hundreds or thousands of results. Paging through those results via interfaces like Figure 5.1 is tedious and inefficient, with no easy way to see trends that cut across documents. By summarizing across a number of documents, rather than just a surrogate for a single video, the user can:

- Be informed of such trends cutting across video documents
- Be shown a quick indicator as to whether the results set, or in general the set of video documents under investigation, satisfies the user's information need
- Be given a navigation tool via the summary to facilitate targeted exploration

As automated processing techniques improve, more metadata is generated with which to build interfaces into the video. For example, all the metadata text for a video derived from speech recognition, overlaid text VOCR processing, and other means can be further processed into people's names, locations, organizations, and time references by extracting named entities. Named entity extraction from broadcast news speech transcripts has been done by MITRE via Alembic [Merlino 1997], and BBN with Nymble [Bikel 1997], and [Miller 1999]. Similarly, Informedia processing starts with training data where all words are tagged as people, organizations, locations, time references, or something else. A tri-gram language model is built from this training data using a statistical language modeling toolkit [Clarkson 1997], which alternates between a named-entity tag and a word, i.e. –person- Rusty –person- Dornin –none- reporting –none- for –organization- CNN –none- in –location- Seattle. To label named entities in new text, a lattice is built from the text where each text word can be preceded by any of the named-entity tags. A Viterbi algorithm then finds the best path through the named-entity options and the text words, just like speech recognition hypothesis decoding.

With a greater volume of metadata describing video, e.g., lists of people's names, locations, etc., there needs to be an overview capability to address the information overload. Prior work in information visualization has offered many solutions for providing summaries across documents and handling volumes of metadata, including:

- Visualization by Example (VIBE), developed to emphasize relationships of result documents to query words [Olsen 1993]
- Scatter plots for low dimensionality relationships, e.g., timelines for emphasizing document attributes mapped to production date [Christel 1999] and [Crane 2001]
- Colored maps, emphasizing geographic distribution of the events covered in video documents [Christel 2000] and [Crane 2001]

Each technique can be supplemented with dynamic query sliders [Ahlberg 1994], allowing ranges to be selected for attributes such as document size, date, query relevance, and geographic reference count.

Consider a query on "air crash" against a 2001 CNN news video subset in the Informedia library. This query produces 998 hits, which can be overviewed using the timeline visualization shown in Figure 5.7. The visualizations shown here convey semantics primarily through positioning, but could be enriched by overlaying other information dimensions through size, shape, and color, as detailed in [Christel 1998] and [Christel 1999]. By dragging a rectangle bounding only some of the green points representing stories, the user can reduce the result set to just those documents for a certain time period and/or relevance range.

For more complex queries with multiple words, the VIBE plot of documents can be used to understand the mapping of results to each term and to navigate to documents that, for example, match the first two words but not a third from the query. VIBE allows users unfamiliar or uncomfortable with Boolean logic to manipulate results based on their query word associations.

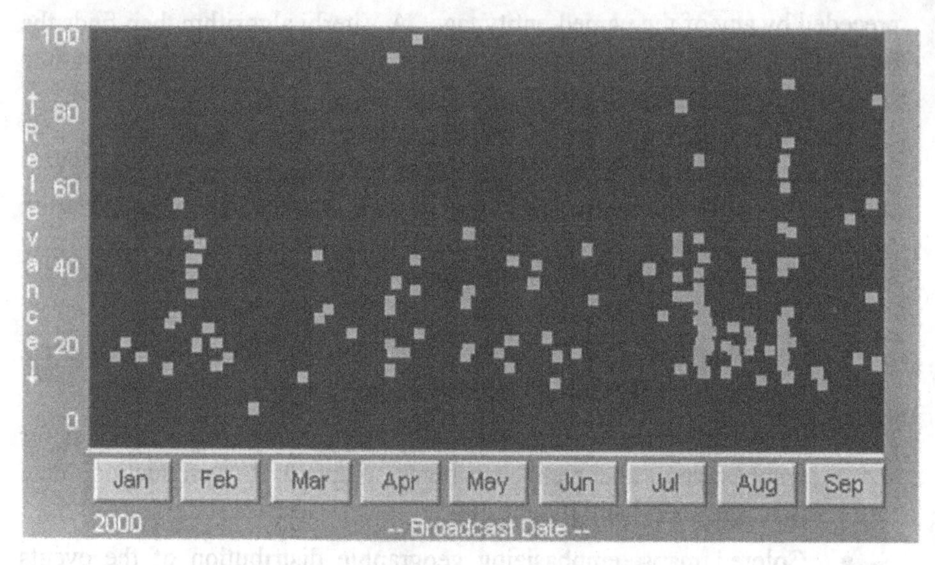

Figure 5.7 Timeline overview of "air crash" query.

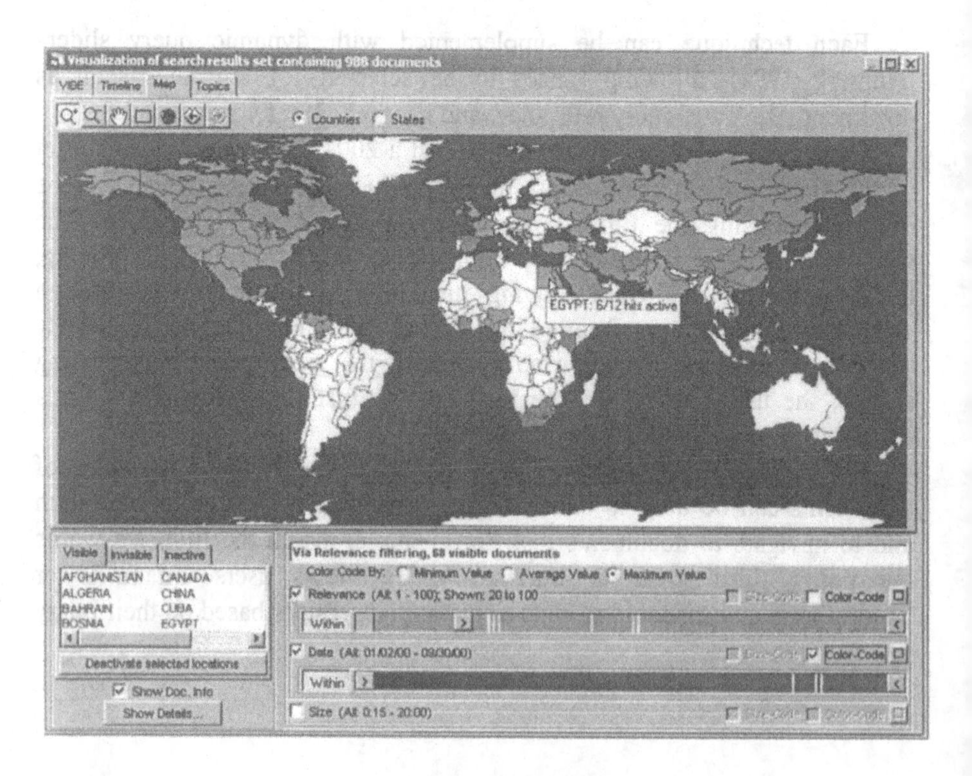

Figure 5.8 Map visualization for results of "air crash" query, with dynamic query sliders for control and feedback.

5.9 Temporal and Spatial Visualizations

For a news corpus, there are other attributes of interest besides keywords, such as time and geography. Figure 5.8 shows both of these attributes in use, as well as the search engine's relevance score. The locations that the video documents refer to are identified automatically through named entity extraction, and stored as the document metadata for use in creating such maps.

Figure 5.8 illustrates the use of direct manipulation techniques to reduce the 998 documents in Figure 5.7 to set of 68 documents under current review. By moving the slider end point icon ≥| with z mouse or pointing device, only those documents having relevance ranking of 20 or higher are left displayed on the map. As the end point changes, so does the number of documents plotted against the map. For example, Brazil, which appeared in documents ranked with relevance score 19 or lower and initially was colored on the map, drops out of the visible, colored set when the higher relevance ranking is specified. Similarly, the user could adjust the right end point for the slider, or set it to a period of, say, one month in length in the date slider and then slide that one month "active" period within January through September 2000 and see immediately how the map animates in accordance with the active month range.

The map is color-coded based on date, and the dynamic sliders show distribution of values based on the country under mouse focus. For example, the 6 "Egypt" stories with relevance > 20 have the relevance and date distribution shown by the yellow stripes on the relevance bars. It has been previously reported that using sliders as a filtering mechanism and showing data distributions can increase efficient use of display space [Eick 1994]. Variations of the geographical interfaces also provide location-based services in mobile devices [Smith 2004] and [Davis 2004].

5.10 Multiple Document Summarization

The visualizations shown in Figure 5.7 and Figure 5.8 do not take advantage of the visual richness of the material in the video library. For the Informedia CNN library, over 1 million shots are identified with an average length of 3.38 seconds, with each shot represented by a thumbnail image as shown in earlier Figures 5.1 through Figure 5.6. Video documents, i.e., single news stories, average 110 seconds in length. This results in an average image count of 32.6 thumbnails for the document storyboards of these stories. These thumbnails can be used instead of points or rectangles in VIBE and timeline plots, and can be overlaid on maps as well. Ongoing research is looking into reducing the number of thumbnails intelligently to produce a more effective visualization [Christel 2002]. For example, by using query-based imagery (keeping only the shots where matches occur) and folding in domain-specific heuristics, the candidate thumbnail set for a visualization can be greatly reduced. With the news genre, a heuristic in use is to remove all studio shots of anchors and weather maps [Christel 2002].

Consider again the 998 documents returned from the "air crash" query shown in earlier figures and densely plotted in Figure 5.7. Through a VIBE plot mapping stories to query words, the user can limit the active set to only the results matching both "air" and "crash". Or, through a world map like Figure 5.8, the user can limit the results to those stories dealing with regions in Africa. The resulting plot of 11 remaining stories and thumbnails is shown in Figure 5.9.

With the increased display area and reduced number of stories in the active set, more information can be shown for each story cluster. Synchronization information kept during the automatic detection of shots for each video, representative images for each shot, and dialogue alignment of spoken words to video can be used to cluster text and images around times within the video stories of interest to the user. For a query result set, the interesting areas are taken to be those sections of the video where query terms ("air crash") are mentioned.

Prior work with surrogates underscores the value of text phrases as well, so text labels for document clusters will likely prove useful. Initial

Figure 5.9 Filtered set of video documents from Figure 5.1, with added annotations.

investigations have shown that displaying common text phrases for documents focused by visualization filters, (e.g., the 11 documents from the full set of 998 remaining in Figure 5.9) communicates additional facts to the user and supplements the visualization plots [Christel 2002]. More extensive evaluation work like that conducted for Informedia single document surrogates will need to be performed in order to determine the utility of video digests for navigating, exploring, and collecting information from news libraries.

Initial interfaces for the Informedia digital video library consisted of surrogates for exploring a single video document without the need to download and play the video data itself. As the library grew, visualization techniques such as maps, timelines, VIBE scatter plots, and dynamic query sliders were incorporated to allow the interactive exploration of sets of documents. "Well-designed human-machine interfaces that combine the intelligence of humans with the speed and power of computers will play a major role in creating a practical compromise between fully manual and completely automatic multimedia information retrieval systems" [Chang 1999].

The interface must provide a view into a video library subset, where the user can easily modify the view to emphasize various features of interest. Summaries across video documents let the user browse the whole result space without having to resort to the time-consuming and frustrating traversal of a large list of documents. The visualization techniques discussed here allow the user efficient, effective direct manipulation to interact with and change the information display.

5.11 Visualization Interface Settings

Modifications to a skim creation interface are necessary for customization. Interfaces for skim creation facilitate efficient testing of the video parameters described in sections 4.4 through 4.6. These interfaces should allow video characterization display and editing for skim rule modifications. They must also allow several display features for skim playback. Several skims were created with various types of video using the Domain Specific Rules described in section 4.7.

The layout and format for skims was also tested. Several alterations to the skim described in Chapter 4 are described for creating improved summarization and user preference. For static representations, or "poster frames", a single frame of each skim scene is used. Experiments include improvements to poster-frame selection, and user control for custom and variable rate skims.

User Interfaces for Skims

An interface for skim experimentation is necessary for efficient testing of the many parameters for skim creation. The interface shown in Figure 5.1 was one such example. With it, a user can automatically view video skims at user specified compaction levels. It can display video characterization data, such as that in Figure 4.8, and is useful for research and development in skim rule creation. It also contains a capability of a fully functional video playback for full-length videos or skims. Processing for individual characterization data is done off-line with several other systems. The video for this interface was digitized and compressed as an MPEG-1 sequence. Earlier interfaces used the SGI Movie format and were designed to take advantage of the video editing and processing tools available on the SGI workstations.

Skim Layout

The formats for skim presentation extend beyond the layout presented in chapter 4. Alterations to the audio and image selection parameters may result in drastically different skims during playback. One alternate representation for the visual portion of a skim is the use of static images, which require less storage and retain the essential visual content. Poster frames may be used at shot breaks and supplemented with additional frames in motion areas to compensate for changes in scenery. Figure 5.3 illustrates an example of the full motion skim and a poster frame video skim. Controls and interfaces for different layouts are used for the skim experimentation and user studies described in chapter 6.

Thumbnail Links to Summaries

In addition to dynamic video representations, the characterization analysis used for selecting important image and audio in skims can be controlled in an interface to select static poster-frames. Poster-frames are used to represent video segments after a query in the Informedia Digital Video Library. An example the Informedia interface with poster-frames was shown in Figure 5.2.

Thumbnail displays can be used as indexes to video content or summaries. The size of the thumbnail can indicate the relative size of the segment. An example of this concept is illustrated in the interface in Figure 5.10, where the thumbnail depth is proportional to the duration of the video segment.

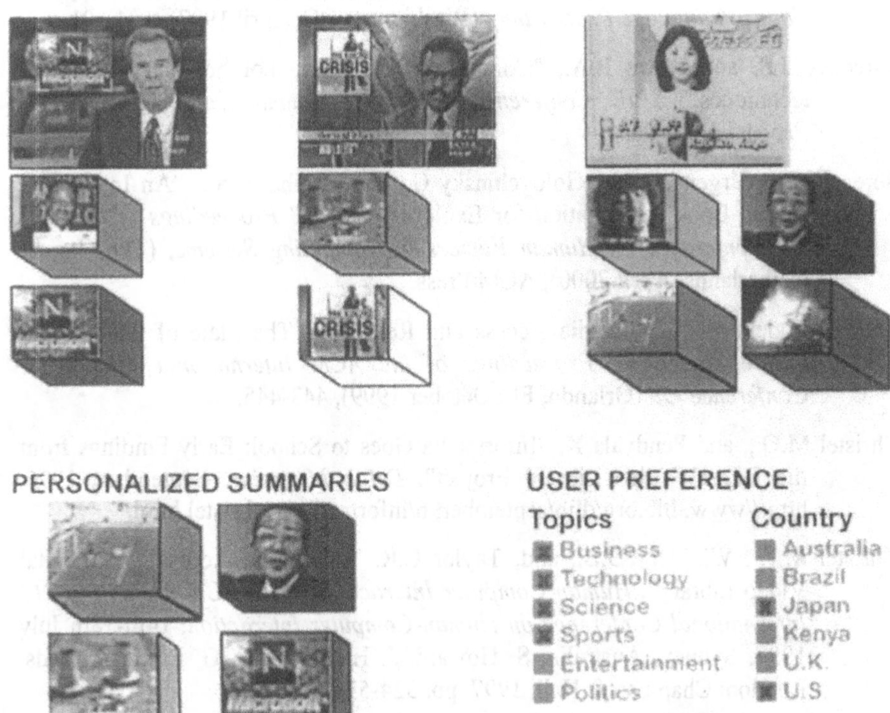

Figure 5.10 Thumbnail links to Video summaries in a customized video retrieval system – Copyright AVA Media Systems.

5.12 Conclusions

Visual interfaces provide functionality and usability in video characterization applications. There are many visualization techniques for summarization, thumbnail display and video library systems. Chapter 6 describes methods to test and evaluate the visualization and summarization techniques described in this chapter.

5.13 Bibliography

Ahlberg, C. and Shneiderman, B. "Visual Information Seeking: Tight Coupling of Dynamic Query Filters with Starfield Displays". *Proceedings of ACM CHI '94* (Boston, MA, April 1994), 313-317.

Bikel, D. M., Miller, S., Schwartz R., and Weischedel R., 1997. Nymble: a high-performance learning name-finder, *Proceedings of the 5th Conf. on Applied Natural Language Processing,* (Washington DC, April 1997), 194-201.

Boreczky J.S. and Rowe L.A., "Comparison of video shot boundary detection techniques." *SPIE Conference on Visual Communication and Image Processing*, 1996.

Boreczky, J., Girgensohn A., Golovchinsky G., and Uchihashi S.. "An Interactive Comic Book Presentation for Exploring Video," *Proceedings of the CHI '00 Conference on Human Factors in Computing Systems*, (The Hague, Netherlands, April 2000), ACM Press.

Chang, Moderator. Multimedia Access and Retrieval: The State of the Art and Future Directions. *Proceedings of the ACM International Multimedia Conference '99* (Orlando, FL, October 1999), 443-445.

Christel M.G.,, and Pendyala K. "Informedia Goes to School: Early Findings from the Digital Video Library Project". *D-Lib Magazine*, September, 1996, http://www.dlib.org/dlib/september96/informedia/09christel.html.

Christel M.G., Winkler D.B., and, Taylor C.R. "Improving Access to a Digital Video Library". *Human-Computer Interaction INTERACT '97: IFIP TC13 International Conference on Human-Computer Interaction*, 14th-18th July 1997, Sydney, Australia, S. Howard, J. Hammond & G. Lindgaard, eds. London: Chapman & Hall, 1997, pp. 524-531.

Christel, M.G., Smith, M.A., Taylor, C.R, and Winkler D.B., "Evolving Video Skims into Useful Multimedia Abstractions". *Proceedings of the CHI '98 Conference on Human Factors in Computing Systems*, C. Karat, A. Lund, J. Coutaz and J. Karat, eds. (Los Angeles, CA, April 1998), pp. 171 - 178.

Christel, M. "Visual Digests for News Video Libraries", *Proceedings of the ACM International Multimedia Conference '99* (Orlando FL, Nov. 1999), ACM Press, pp. 303-311.

Christel M.G. and Warmack A.S., "The Effect of Text in Storyboards for Video Navigation". *Proceedings IEEE International Conference on Acoustics, Speech, and Signal Processing (ICASSP),* Salt Lake City, UT, May 2001, Vol. III, pp. 1409-1412.

Christel M., Olligschlaeger, A.M, and Huang, C., "Interactive Maps for a Digital Video Library". *IEEE Multimedia* 7(1), 2000, 60-67.

Christel, M., and Martin, D. "Information Visualization within a Digital Video Library". *Journal of Intelligent Information Systems* 11(3) (1998), 235-257.

Christel, M., Hauptmann, A., Wactlar, H., and Ng, T., "Collages as Dynamic Summaries for News Video", *Proceedings of the ACM International Multimedia Conference '02* (Juan-les-Pins, France, Dec. 2002), ACM Press.

Clarkson, P. and Rosenfeld, R., "Statistical language modeling using the CMU-Cambridge toolkit", in *Proc. Eurospeech '97* (Rhodes, Greece, Sept. 1997), International Speech Communication Assoc., 2707-2710.

Crane, G., Chavez, R., et al. "Drudgery and Deep Thought". *Communications of the ACM* 44(5), 2001, 34-40.

Davis, M., and Sarvas R., "Mobile Media Metadata for Mobile Imaging." *Proceedings of IEEE International Conference on Multimedia and Expo* (ICME 2004) Special Session on Mobile Imaging in Taipei, Taiwan.

Eick, S.G. "Data Visualization Sliders". *Proceedings of the ACM Symposium on User Interface Software and Technology* (Marina del Rey, CA, Nov. 1994), ACM Press, 119-120.

Firmin T. and, Chrzanowski M.J. "An Evaluation of Automatic Text Summarization Systems", in Maybury, M.T. ed., *Advances in Automatic Text Summarization*, The MIT Press, Cambridge, MA, 1999

Hampapur, A. Jain, R. and Weymouth, T. "Production Model Based Digital Video Segmentation", *Multimedia Tools and Applications*, **1**, 1995, pp. 9-46.

Hearst, M. A. "TileBars: Visualization of Term Distribution Information in Full Text Information Access", *Proceedings of the ACM CHI'95 Conference on Human Factors in Computing Systems*, Denver, CO, May, 1995 59-66.

Hegarty, M. and Just, M.A. "Constructing mental models of machines from text and diagrams". J. Memory & Language, Dec. 1993, 32(6):717 – 742.

Komlodi A. and Slaughter, L. "Visual Video Browsing Interfaces Using Key Frames", *Proceedings of the ACM CHI'98 Conference on Human Factors in Computing Systems - Summary*, ACM, New York, 1998, pp. 337-338.

Li, F., Gupta, A., Sanocki, E., He, L., and Rui, Y. "Browsing Digital Video", *in Proceedings of the ACM CHI'00 Conference on Human Factors in Computing Systems,* The Hague, Netherlands, April 2000, ACM Press.

Large, A., Beheshti, J., Breuleux, A., and Renaud, A. "Multimedia and comprehension: The relationship among text, animation, and captions". *Journal of American Society for Information Science*, June 1995.

Lienhart R. et al. "Video Abstracting," *Communications of the ACM, 40*, 12, pp. 54-62, 1997.

Mauldin, M. "Information Retrieval by Text Skimming", PhD Thesis, Carnegie Mellon University. August, 1989 (also available as CMU Computer Science technical report CMU-CS-89-193).

Merlino, A., Morey, D., and Maybury, M. "Broadcast News Navigation using Story Segmentation", *Proceedings of the ACM International Multimedia Conference '97* (Seattle WA, Nov. 1997), ACM Press, 381-391.

Miller, D., Schwartz, R., Weischedel, R., and Stone, R. "Named Entity Extraction for Broadcast News", in *Proceedings DARPA Broadcast News Workshop*

(Washington DC, March 1999), http://www.nist.gov/speech/publications/darpa99/html/ie20/ie20.htm.

Nielsen, J., and Mack, (R.L. eds.). "Usability Inspection Methods". John Wiley & Sons, New York, NY, 1994.

Nugent, G.C. "Deaf students' learning from captioned instruction: The relationship between the visual and caption display". *Journal of Special Education*, 1983, 17(2):227 – 234.

Olsen, K.A., Korfhage, R.R., et al. "Visualization of a Document Collection: The VIBE System". *Information Processing & Management* 29(1) (1993), pp. 69-81.

Rowley, H., Kanade, T., and Baluja, S., "Neural Network-Based Face Detection ", *IEEE Transactions on Pattern Analysis and Machine Intelligence*, January 1998.

Sato, T. Kanade, T., Hughes, E. Smith, M. "Video OCR for Digital News Archive," *IEEE International Workshop on Content-based Access of Image and Video Databases, ICCV*, Bombay, India, January 1998.

Smith, M., "Integration of Image, Audio, and Language Technology for Video Characterization and Variable-Rate Skimming", PhD Thesis, Carnegie Mellon University. January, 1998.

Smith, M., Choi, A., Aublant, S., "Mobile Image Capture and Management", *Joint Conference on Digital Libraries* (JCDL), Demonstration, Tuscon, AZ, June 7-11, 2004.

Schneiderman, H., and Kanade, T., "Object Detection Using the Statistics of Parts ", *International Journal of Computer Vision*, 2002.

Taniguchi, Y., Akutsu, A., Tonomura, Y., and Hamada, H., "An Intuitive and Efficient Access Interface to Real-Time Incoming Video Based on Automatic Indexing", *Proceedings of the ACM International Multimedia Conference*, ACM, New York, 1997, pp. 25-33.

Uchihashi, S., Foote, J., Girgensohn, A., and Boreczky. J., "Video Manga: Generating Semantically Meaningful Video Summaries", *Proceedings of the ACM International Multimedia Conference*, (Orlando, FL) ACM Press, pp. 383-392, 1999.

Wactlar, H., Christel, M., Gong, Y., and Hauptmann, A. "Lessons Learned from the Creation and Deployment of a Terabyte Digital Video Library". *IEEE Computer* 32(2) (Feb. 1999), 66-73.

Yeo, B.-L. and Yeung, M.M. "Retrieving and Visualizing Video", *Communications of the ACM* 40 (12), 1997, pp. 43-52.

Zhang, H.-J., Smoliar, S.W., Wu, J.H., Low, C.Y., and Kankanhalli, A., "A Video Database System for Digital Libraries," *Lecture Notes in Computer Science* 916, 1995, pp. 253-264.

Chapter 6
Evaluation

Image and audio understanding technology is tested with quantitative measures. Evaluating the output of a video summarization system is subjective in nature and difficult to analyze with objective experimentation. Extensive subjective testing is used to assess an accuracy or confidence measure.

User studies provide the best quality measure for video characterization. The interface and visual appeal are important factors in determining the success of applications in summarization and segmentation. This chapter presents a description of subjective evaluation methods and user studies for video summaries. The experimental setup and results from two video skim user studies and a comparative study of poster frames and moving skims are presented.

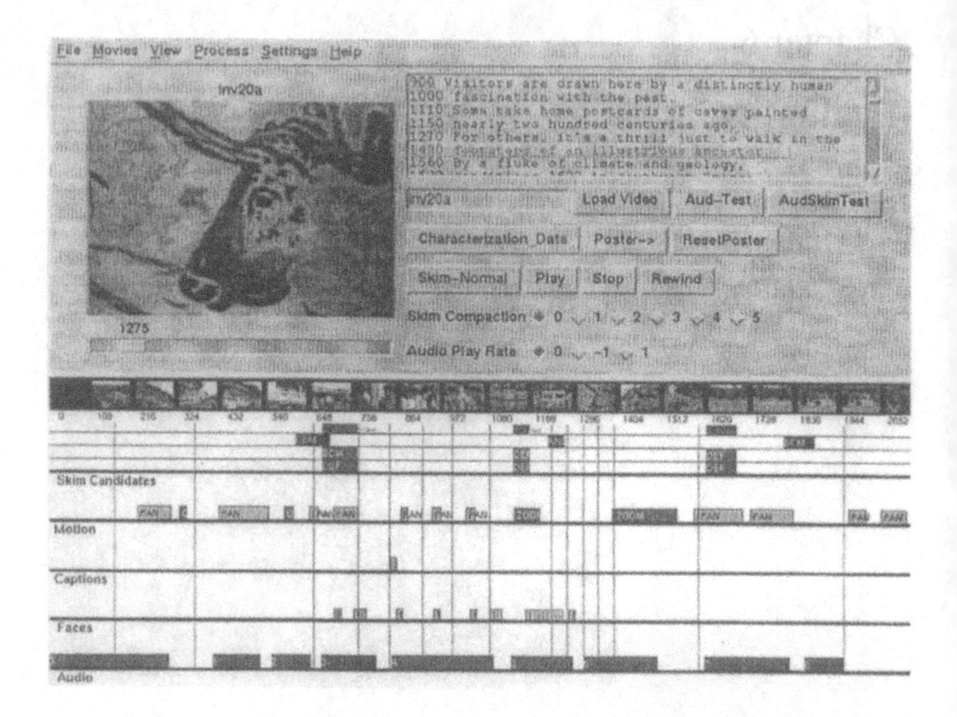

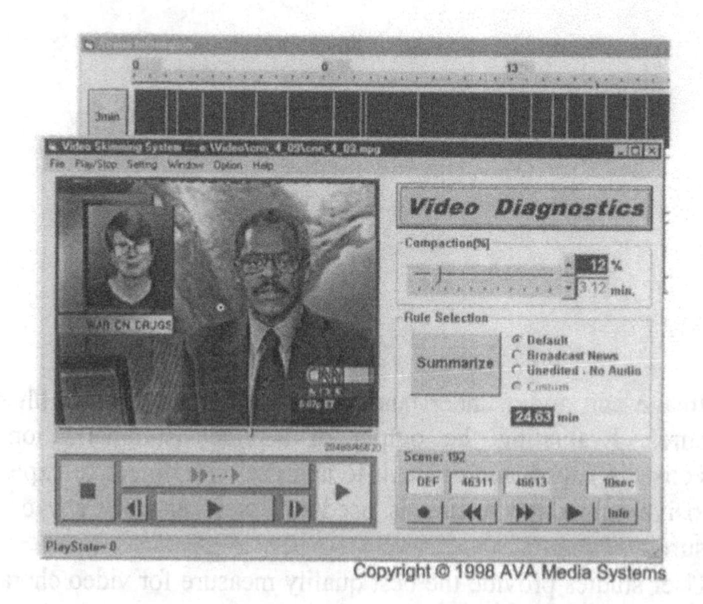

Figure 6.1 Experimental interfaces for implementing video summaries: (Top) experimental system for testing video summaries, playback speed, compaction settings, and forward and reverse play; (Bottom) System Interface of AVA Media Systems and OGIS Research Institute, Osaka Japan, for video characterization and hierarchical video summary rule analysis.

6.1 Experimentation and Evaluation

Video characterization and summarization requires subjective experimentation. Interfaces for testing summary creation must facilitate efficient creation and customization of video rules and parameters, as discussed in chapter 5. Figure 6.1 shows two experimental systems for video characterization and summarization analysis. The interfaces display video features and provide editing tools for modifying skim rules. They also provide display functions for conventional video playback, such as fast forward, single frame advancement and a temporal scroll bar. These interfaces were used to create skims using various types of video and the domain specific rules described in section 4.7.

Two studies are presented in sections 6.3 and 6.4 to measure the effects of different dynamic skim techniques on comprehension, navigation, and user satisfaction. Video skim layout and formatting requires subjective experimentation. The automated skim and other visualization schemes described in chapters 4 and 5 can be altered to create custom summarization applications. Results from an initial study helped refine video skims, which were then assessed in the second experiment. Significant benefits were found for skims built from audio sequences that were extracted using the audio descriptors from chapters 3 and 4. The experimentation for user-studies with these skims is described in later sections.

A third study is presented in section 6.5 for comparison of static and dynamic summarizations. In addition to testing visual layout, user-studies evaluate the summarization and browsing utility of the skim. They include experimentation with poster frames in digital video libraries. For these static representations, a single frame of each scene or selected skim scene is used. These experiments improve poster-frame selection and assess user response to static and dynamic summaries.

6.2 Skims used for User Studies

User studies assess quality when subjective measures are needed for experimentation. Language understanding systems employ user studies in areas such as text abstraction and topic identification. There is also a large area of subjective research in image and video quality assessment [Pinson 2003]. The skims described in this section are generated automatically and their quality and effectiveness are tested against the original video sequences. The testing process focuses on evaluating the selected and concatenated video and audio portions of the summary. Audio distortion, frame rate and other video artifacts should also be considered in a summarization evaluation.

Distortion at Different Frame Rates

One simple approach to creating skims is to change the frame rate across the entire video. The result will achieve a decrease in viewing time, but it will also perturb the audio [Degen 1992] and distort image information. Maintaining a coherent presentation is thus unlikely using only fast playback [Stevens 1992]. Degen and other researchers have developed pitch-controlled methods for accelerating audio and video. Audio intelligibility is maintained at rates up to 2x the original rate. Video summaries should preserve the frame rate of the original video or use some pitch-controlled algorithms to avoid audio distortion. The user studies in sections 6.3 and 6.4 use skims played at their original frame rate.

Fixed Intervals

Subsampling is another simple method for summarization that simply skips video frames at fixed intervals, for example, displaying the first 10 seconds at 100 second intervals. Dropping video at regular intervals will likely delete essential information [Wactlar 1996], but this technique is trivial to implement and so serves as a default skim (DEF) in the user studies reported later.

Skim Formats

Skim generation can be derived from individual or combined techniques in image, audio, or language processing at differing degrees. In this section, we describe four types of skims created for testing in the first user-study. The formats are a Default skim, an Image-centric skim, an Audio-centric skim, and an Integrated skim that combines audio and image characterization technology.

An image-centric skim (IMG) emphasizes image information. After decomposing the source into component shots [Hampapur 1995], [Pfeiffer 1996], [Taniguchi 1995], [Yeung 1995], and [Zhang 1995], image heuristics, including weighting highly those frame sequences with significant camera motion and those showing people or a text caption, prioritize shots for inclusion in the skim. Super-Rules avoid overlapping regions and provide even coverage throughout the source video. Sections of high-value video are repeatedly added in this manner until the skim reaches a threshold size, such as one-tenth the duration of the full video.

An audio-centric skim (AUD) concentrates on audio information. Automatic speech recognition and alignment techniques [Hauptmann 1997] align the audio track tightly to the word transcript. The link-grammar parser developed at Carnegie Mellon University identifies noun phrases within the transcript, and term-frequency inverse-document-frequency (TF-IDF) analysis yields a ranking of noun phrases [Mauldin 1991]. Words that appear often in a particular document but relatively infrequently in a standard corpus receive the highest weight. Noun phrases with many such words are judged "keyphrases," and the part of the source video that corresponds to keyphrases is considered the most important information. Video containing keyphrases is repeatedly added until the skim reaches a threshold size.

An "integrated best" skim (BOTH) combines the image-centric and audio-centric skim approaches. Top-rated audio regions are augmented with video selected from a larger window, extending five seconds before and after the audio region. The "best video" image heuristics of IMG select portions of this expanded video window for inclusion. While the resulting audio and video components may not be precisely synchronized, each captures the most significant information of its type.

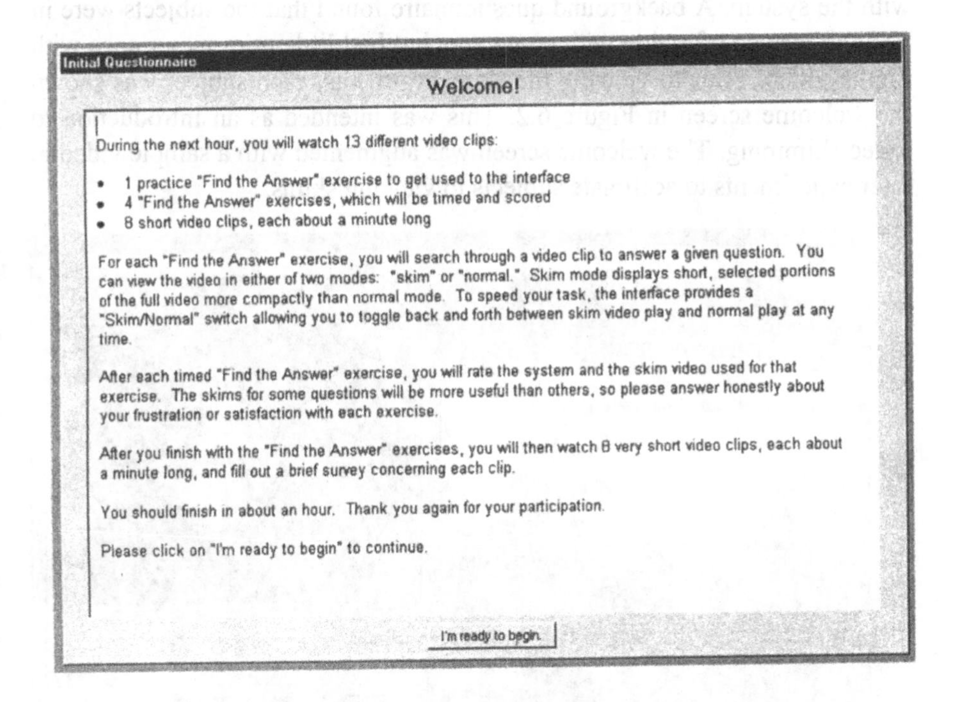

Figure 6.2 Welcome screen for fact-finding and gisting skim study - experiment I. Video skims appear accelerated during first glance, which imposes a negative bias on the initial skim treatment. In the subsequent studies, users were shown a sample video summary to acclimate them to the skim viewing experience.

6.3 Skim Study - Experiment I

The first experiment examines the use of four different types of video skims for two information retrieval tasks:

- Fact-finding – Users attempt to locate the portion of the video that matches the question using a specified skim treatment. An example of the interface for the fact-finding exercise is shown in Figure 6.3.
- Gisting - Users attempt to match the image and text phrases that belong to the original video based on viewing a skim treatment. An example interface for the gisting exercise is shown in Figure 6.4.

Subjects

Forty-eight Carnegie Mellon University students volunteered for the study in the spring of 1997. Each received $5 for spending about an hour with the system. A background questionnaire found that the subjects were in general very comfortable with computers but had little prior experience with digital video. Prior to viewing the skim treatments, each subject was shown the welcome screen in Figure 6.2. This was intended as an introduction to video skimming. The welcome screen was augmented with a sample video in later experiments to acclimate subjects to viewing skims.

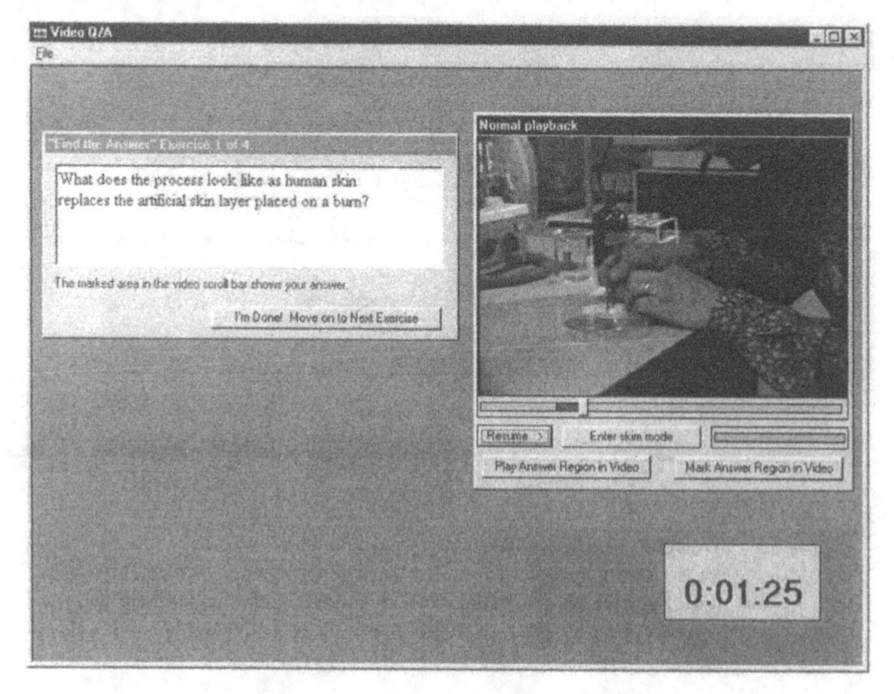

Figure 6.3 Fact-finding interface for skim study - experiment I. Users attempt to locate the portion of the video that matches the question using a specified skim treatment.

Design

Each of the skims was one-tenth the length of the source video and built from segments averaging three seconds in duration, as shown in Figure 6.5. This 3-second grain-size equals the average duration of key phrases used in the AUD and BOTH skims. The study compared the following four types of skims:

DEF Default, simple synchronized skim: subsampled from seconds 1-3 of the source video, then seconds 31-33, seconds 61-63, 91-93, 121-123, etc.

IMG Shown as "best video" skim. The audio is the same as the DEF skim and image selection is based on the highest-ranking image scenes.

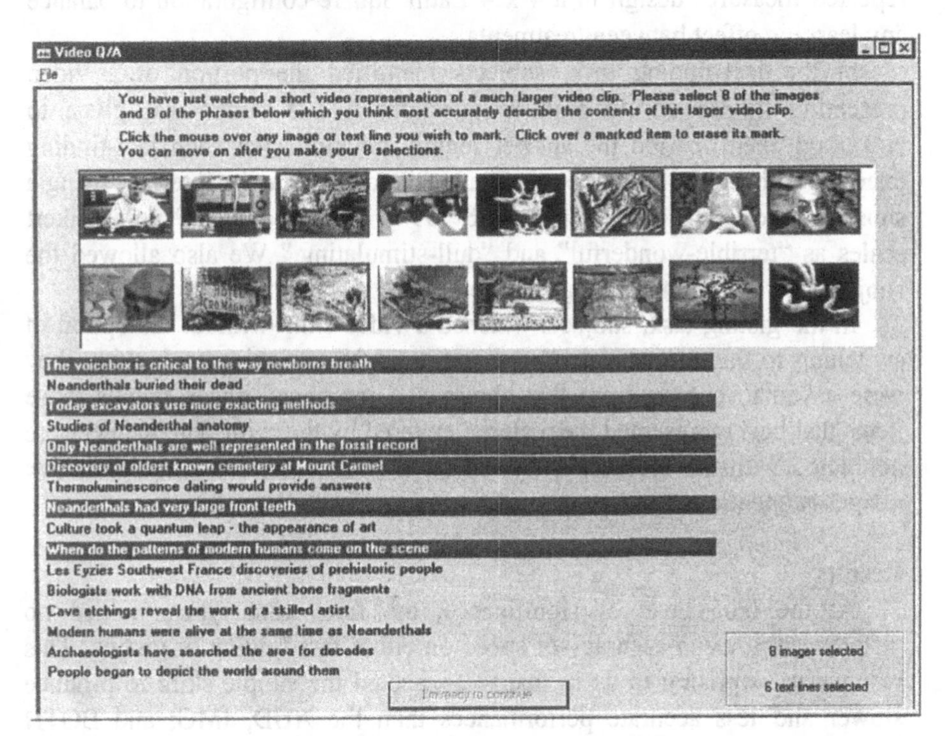

Figure 6.4 Gisting interface for skim study - experiment I. Users attempt to match the image and text phrases that belong to the original video based on viewing a skim treatment. It was found that this type of interface does not provide sufficient separation for individual skim treatments. Users are able to estimate the topic of the video by visual and textual correlation between the choices provided and any of the skim treatments.

AUD Shown as "best audio" skim where selection is based on the highest-ranking audio phrases. The images are synchronized with the audio.

BOTH Shown as "integrated best" skim where the audio is the same as the AUD skim and image selection is based on the highest-ranking image scenes.

Procedure

Subjects participated in the study individually. Each used a computer with a 17-inch color monitor, keyboard, mouse, and headphones. Each subject completed the fact-finding task four times, once for each skim type, and the gisting task eight times, viewing each skim type twice. We used a repeated measures design in a 4 x 4 Latin Square configuration to balance any learning effect between treatments.

In the fact-finding task, subjects identified the portion of a video presenting the answer to a given question. A $25 bonus was given to encourage them to find the answer region quickly. After each fact-finding exercise with a skim type, we asked subjects to evaluate the interface using a subset of the QUIS instrument [QUIS 1994], including such 9 point Likert scales as "terrible-wonderful" and "dull-stimulating." We also allowed the subjects the opportunity to type in open-ended comments.

In the gisting task, subjects watched a video skim without the option of switching to the normal video presentation. After watching each skim, they were asked to choose from text-phrase and thumbnail-image menus those items that best represented the material covered by the skim. The menus were populated with the independently validated text phrases and representative images as mentioned earlier.

Results

At the 0.05 level of significance, the four skim types yielded no differences in mean accuracy or speed on either fact-finding or gisting. This result was surprising to us in that we expected the simple skim to produce slower and less accurate performances than the AUD, IMG, and BOTH skims. Pilot studies had shown us that users found the simple skim "jerky", "too hard to follow", and "too jumpy."

There were also no significant (0.05 level) differences between the QUIS answers concerning user satisfaction for the four skim types.

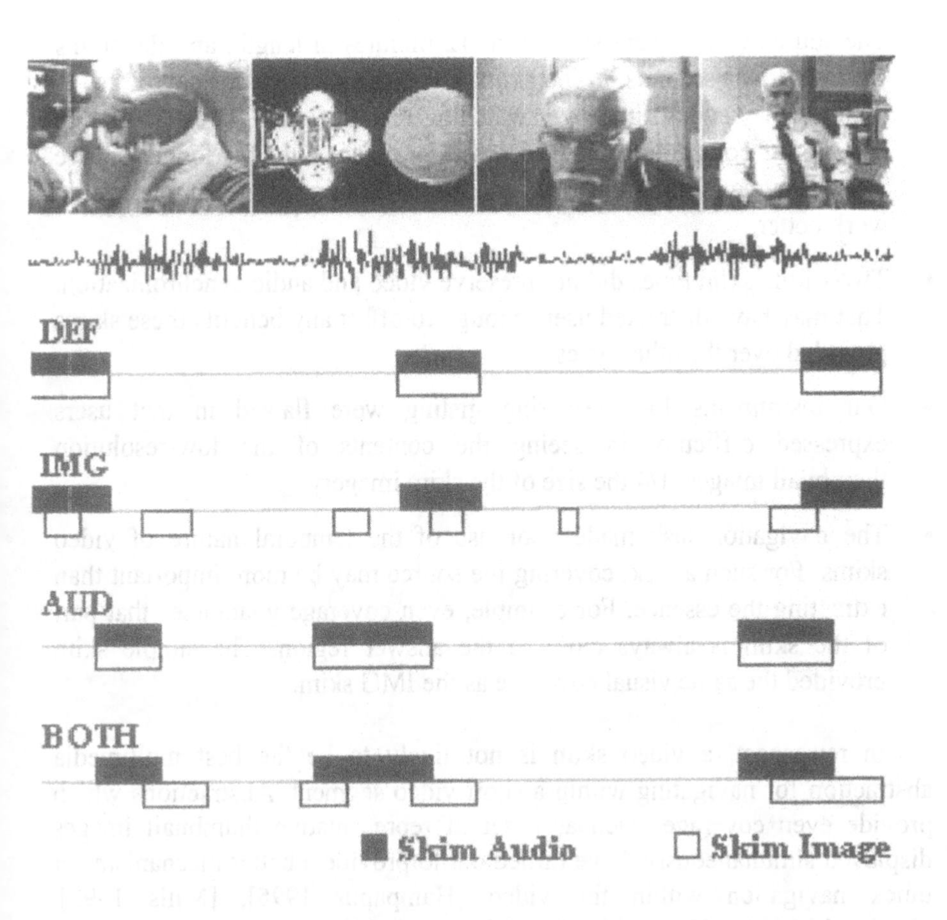

Figure 6.5 Audio/video alignment in four skim designs (*Experiment* I): (DEF) shows a fixed interval treatment, (IMG) shows a fixed interval audio with an automated skim image selection, (AUD) shows an automated audio skim selection with synchronized images, and (BOTH) automated image and audio skim selection.

Redesigning Skim Creation

Several factors may have contributed to the fact that subjects did not observe many differences between the skim types:

- All the video skims used a small grain size. While subjects shown the AUD, IMG, and BOTH skims succeeded in identifying and including "important" segments, the segment durations may have been too short to communicate content effectively. Thus a small grain size may mask any differences among skim designs, leading subjects to notice that all skims were, among others, "jerky" and "too jumpy."

- The source videos were short, 8 to 12 minutes in length, and the skims were 48 to 72 seconds. The skims for these short videos save 7 to 11 minutes of viewing time over watching the full video, but perhaps the benefits of compaction may become significant only for longer source video. Maybe 30-minute videos and 3-minute skims, for example, would work better.
- Two of the skim types did not preserve video and audio synchronization. That may have distracted users enough to offset any benefits these skims provided over the other types
- Our instruments for measuring gisting were flawed in that users expressed difficulty in seeing the contents of the low-resolution thumbnail images, 1/4 the size of the skim imagery.
- The navigation task made poor use of the temporal nature of video skims. For such a task, covering the source may be more important than extracting the essence. For example, even coverage guarantees that part of the skim is always close to the answer region. The simple skim provided the same visual coverage as the IMG skim.

In retrospect, a video skim is not likely to be the best multimedia abstraction for navigating within a short video segment. Abstractions which provide even coverage, such as a set of representative thumbnail images displayed simultaneously, have turned out to provide a better mechanism for quick navigation within the video [Hampapur 1995], [Mills 1992], [Taniguchi 1995], [Yeung 1995], and [Zhang 1995]. Showing the location of matching query words within the video's transcript also aids navigation to a point within a video more directly than a skim [Christel 1997]. Rather than attempting to justify the use of skims for navigation, we decided to address only the issue of gisting in our subsequent skim study.

6.4 Skim Study - Experiment II

We addressed the shortcomings of the those skim types used in the first study and the gisting experiment by:

- Modifying the key phrase selection process so that the average grain size would now be 5 seconds rather than 3 seconds.
- Creating skims of half-hour videos.
- Decreasing the synchronization offsets for the audio and video in the BOTH "integrated best" skim.

- Improving the instruments for measuring gist, including the use of images with the same resolution as the video (352 x 240 pixels for MPEG-I video).

We pilot-tested the new generation skims from which we obtained the following feedback:

- People question how good these skims are compared to the full video.
- People were more annoyed by choppy audio than choppy video.
- Some people took the extremist position that the audio carries all the of the gist for a movie and that two skims with the same audio content will produce similar results regardless of their video content.

This feedback directly influenced the design of the subsequent skim study conducted in September 1997. Our main concern with the skims generated for the first study was the grain size. The open-ended text feedback we collected in that study reported that all of the skims were perceived as too disjointed and that the audio in particular was too choppy for easy comprehension.

As an alternate approach to relying solely on transcript analysis, for our second study we grouped words into phrases using signal power to identify breaks between utterances. To detect these breaks we used the audio segmentation techniques discussed in section 3.3. Other researchers have used similar speech signal characteristics to produce compact audio representations [Arons 1993]. Thus, the audio signal itself is used to delineate starts and ends of utterances better. In addition, these phrases are typically longer than the parsed noun phrases used in the first study.

An example of the improved audio segmentation is shown in Figure 6.6, where the line "The sense of taste is probably the sense" is a partial sentence

A- Grammar Parsed Phrase
"The sense of taste is probably the sense"

B - Audio Segmented Phrase
"The sense of taste is probably the sense that is most involved in immediate pleasure"

Figure 6.6 Parsed audio regions: (A) shows the region used in the first experiment obtained by a parsed noun phrase using a Grammar Parser; (B) shows the region used in the second experiment obtained by audio segmentation.

as extracted by the first study. Phrase segmentation based on audio breaks collects the full sentence as a candidate audio region. Phrases are delineated by audio breaks and scored using TF-IDF weighting as before. The highest scoring phrases were included in the skim and averaged five seconds. Skims produced from these longer key phrases were less truncated and less choppy. Detected utterances shorter than eight seconds were included unchanged and only extremely long utterances broken.

Another concern with the skims from the first study was the synchronization between images and audio in the image-centric IMG and integrated BOTH skims. For a given audio phrase, these skims selected images from a considerably broader search window than in the synchronized video. For our second study, we limited the image adjustments to video within the scene of the audio regions near shot breaks or blank video frames.

Figure 6.7 shows four of the skims used in this second experiment. Two subsampled skims DFS and DFL were included in the test to better understand the effects of grain size. The "new integrated best" skim, NEW, used only minor image adjustments to better preserve synchronization between its audio skim. DFL used the same grain size as NEW's average grain size (about 5 seconds) with DFS using half that grain size to test the effects of finer grained skims. The fourth skim, referred to as RND, was included as a bad skim example. It used the same audio as the NEW skim but reversed the video, so that the audio and video were only in alignment at the middle of the skim. The audio played from start to end as with the NEW skim, but the video played from the end to the start.

In summary, the second study had five experimental treatments: four types of skims, each decreasing the size of the associated video by a factor of 7.5, and a fifth treatment which was the full associated video. This compaction captured approximately 8 seconds per minute of video, which is the maximum audio duration for a phrase in the improved audio selection.

Subjects

Twenty-five Carnegie Mellon University students volunteered for the study, and each received $7 for spending about eighty minutes with the system. A background questionnaire revealed that the subjects were in general very comfortable with computers but had little prior experience with digital video.

Materials

The video material was drawn from the three public television series as used in the first study, with manually chosen representative images and text phrases again serving as the gist of a video.

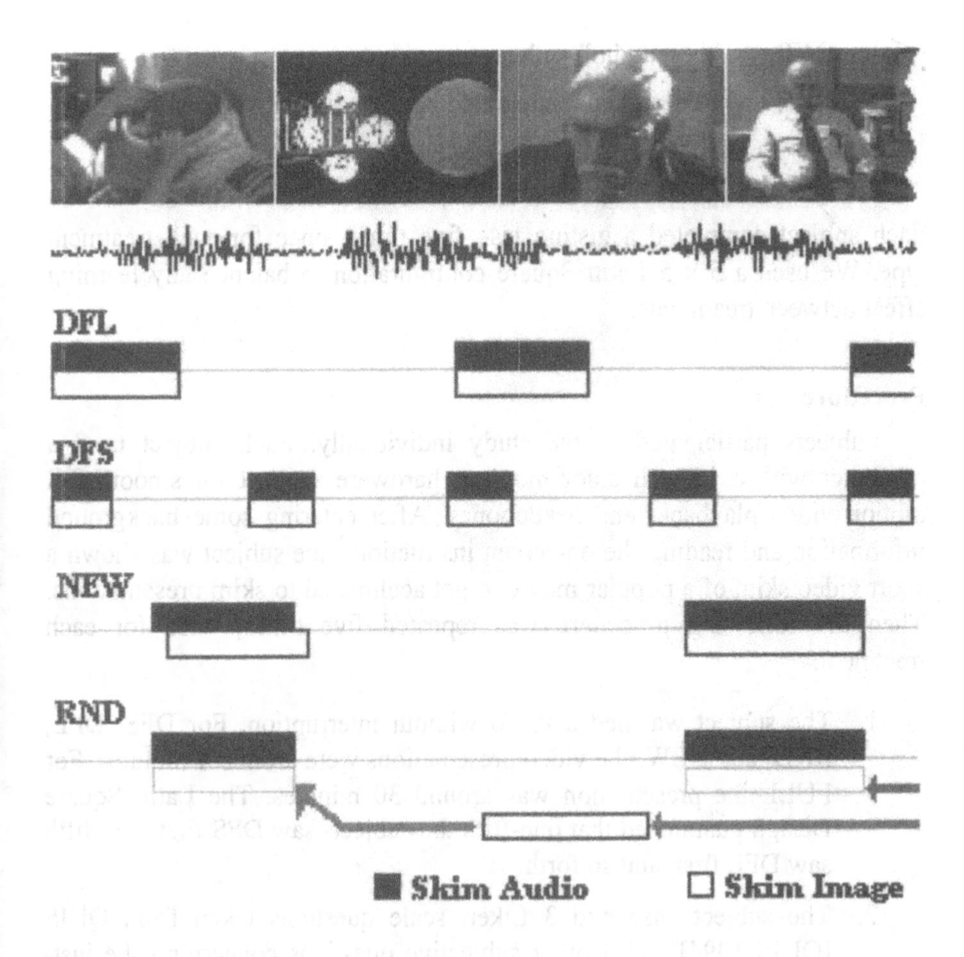

Figure 6.7 Audio/video alignment in four skim designs (Experiment II): (DFL) shows a fixed interval treatment with a longer grain size, (DFS) shows a fixed interval treatment with a small grain size, (NEW) shows an automated audio and image skim selection with close synchronization between audio and imagery, and (RND) is the obvious worst case example with the audio from the NEW skim and a reversed image track from the New image skim.

Design

The five treatments are labeled as follows:

DFS Simple, short skim consisting of seconds 1-2.5 from the full source video, then seconds 21-22.5, seconds 41-42.5, etc.

DFL Simple, long skim consisting of seconds 1-5, then seconds 41-45, seconds 81-85, etc.

RND "best audio" with reordered video

NEW Redesigned, "integrated best" audio and video skim approach

FULL Full source video with no information excluded

Each subject completed a gisting task five times, once for each treatment type. We used a 5 x 5 Latin Square configuration to balance any learning effect between treatments.

Procedure

Subjects participated in the study individually. Each subject used a computer with a 17-inch color monitor, hardware support for smooth full motion video playback, and headphones. After entering some background information and reading the on-screen instructions, the subject was shown a short video skim of a popular movie to get acclimated to skim presentations. Then the following procedure was repeated five times, once for each treatment:

1. The subject watched a video without interruption. For DFS, DFL, RND, and NEW, the video presentations were around 4 minutes. For FULL the presentation was around 30 minutes. The Latin Square Design guaranteed that one-fifth the subjects saw DFS first, one-fifth saw DFL first, and so forth.
2. The subject answered 3 Likert scale questions taken from QUIS [QUIS 1994] and 3 other subjective questions concerning the just-completed video. An example of this subjective question interface is shown in Figure 6.8.
3. The subject was presented with 10 images, one at a time. The images were the same resolution as the video. For each image, the subject had to select "yes" or "no" based on their recall of whether the image was included in the video they just saw. The interface for the new image recall test is shown in Figure 6.9.
4. The subject was presented with 15 text phrases, shown one at a time. For each text phrase, the subject had to select "yes" or "no" based on their interpretation of whether that text phrase summarizes information, which would be part of the full source video. The interface for the new text recall test is shown in Figure 6.10.

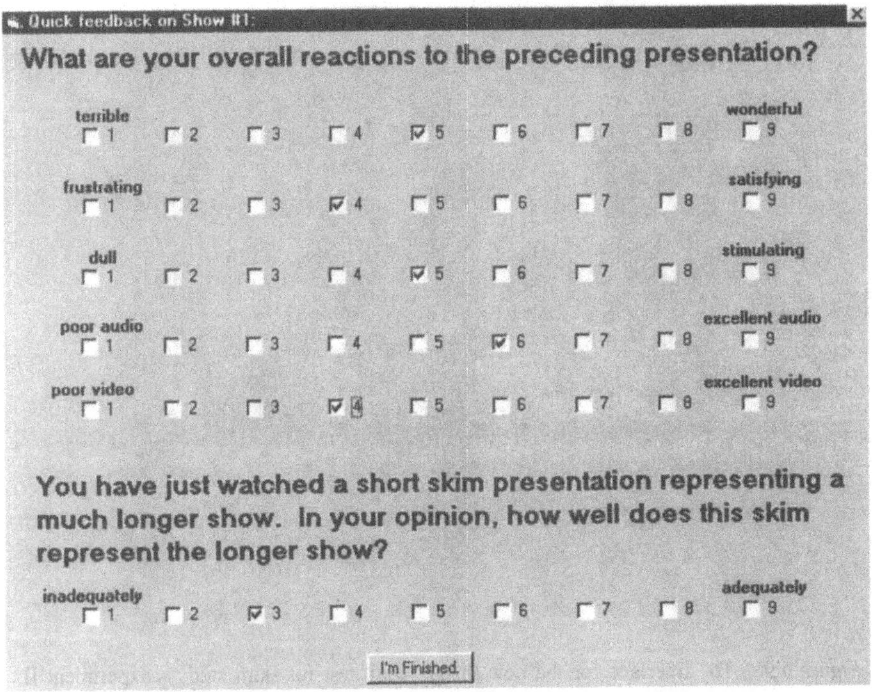

Quick feedback on Show #1:

What are your overall reactions to the preceding presentation?

terrible 1 2 3 4 5 6 7 8 9 wonderful

frustrating 1 2 3 4 5 6 7 8 9 satisfying

dull 1 2 3 4 5 6 7 8 9 stimulating

poor audio 1 2 3 4 5 6 7 8 9 excellent audio

poor video 1 2 3 4 5 6 7 8 9 excellent video

You have just watched a short skim presentation representing a much longer show. In your opinion, how well does this skim represent the longer show?

inadequately 1 2 3 4 5 6 7 8 9 adequately

I'm Finished

Figure 6.8 QUIS subjective questions with 3 Likert scale questions and 3 subjective questions related to audio and image quality and how well the skim treatment represented the original program.

5. The subject was asked how well the video prepared him or her for the just-completed image and text questions. The subject was also given the opportunity to type any open-ended comments concerning this particular video treatment.

User Study Statistics

Analysis revealed highly significant ($p < 0.01$) differences in mean performance on text gisting and image recall among the five video treatments. The Student-Newman-Keuls test (SNK) was subsequently run to determine whether the differences between specific means were significant, enabling us to draw conclusions on the relative merits of the different skim treatments and the "FULL" full source video treatment.

Figure 6.9 The interface for the new image recall test for skim study - experiment II. Users were shown a single image at a time as not to provide clues to the video content. Images were presented at full resolution (352x240).

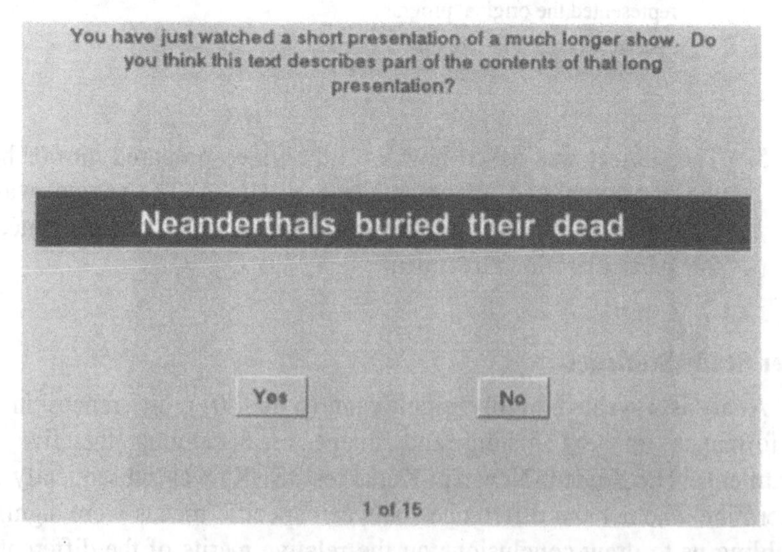

Figure 6.10 The interface for the new text recall test for skim study - experiment II. Users were shown a single phrase at a time as not to provide semantic clues to the video content. Text phrases were displayed in the black field above.

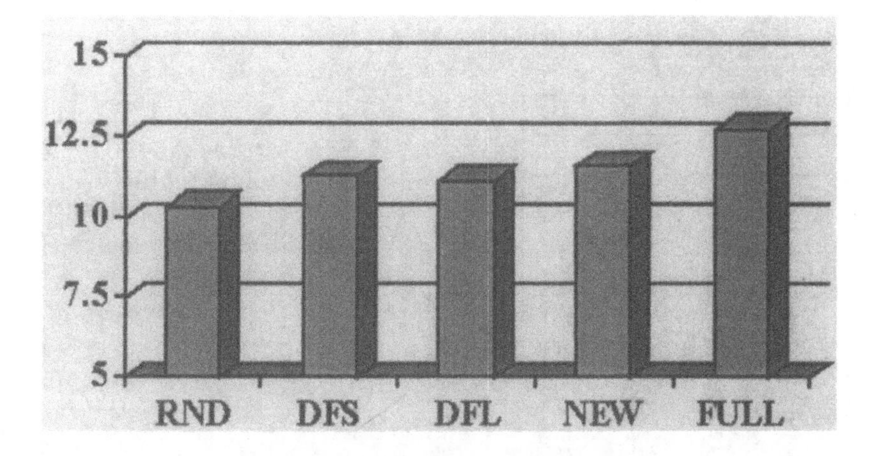

Figure 6.11 Mean scores for image recall.

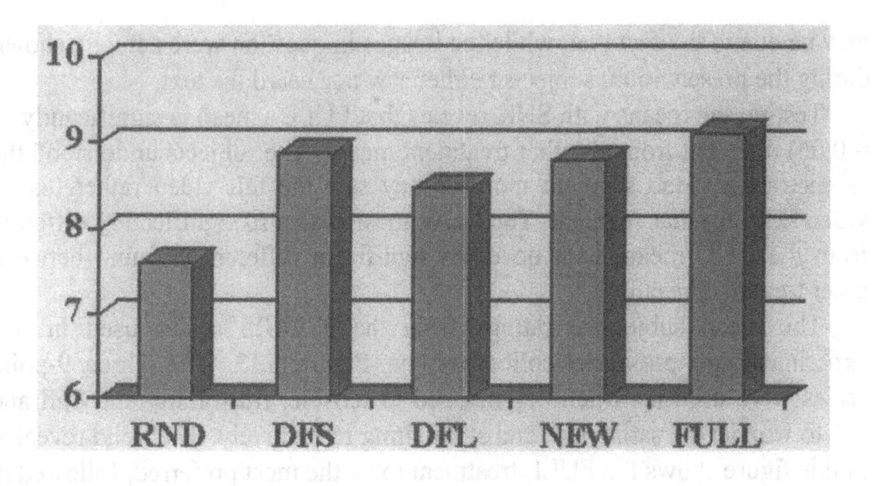

Figure 6.12 Mean scores for text phrase identification.

The mean performance on the 10 image questions is given in Figure 6.11. Testing the means with SNK reveals that RND's mean is significantly ($p = 0.05$) different from all other treatment means. No other significant differences were found between the treatment means, i.e., the other skim treatments were as good at promoting image recall as the full video (FULL). Only when synchronization was extremely poor (the RND treatment) did image recall performance drop off significantly.

The mean performance for the 15 text questions, given in Figure 6.12, was generally worse than that for the image recall questions. This difference

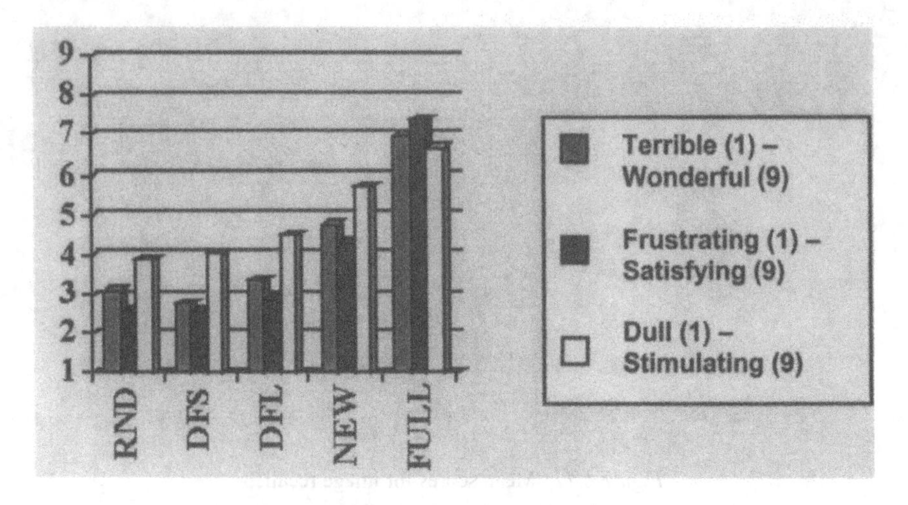

Figure 6.13 Mean scores for 3 QUIS subjective ratings.

may be due to the fact that, while the images in question were actually shown during the presentation, subjects neither saw nor heard the text.

Testing the means with SNK reveals that FULL's mean is significantly (p = 0.05) different from all other treatment means. The subjects understood the essence of a video segment more if they saw the full video rather than a video skim for that segment. The NEW mean was also significantly different from the RND mean, with no other significant differences found between other treatment means.

The mean subjective ratings from the 3 QUIS scales used in this experiment are presented collectively in Figure 6.13. The Three 9-point scales were used in which "1" mapped to terrible, frustrating, and dull and "9" to wonderful, satisfying, and stimulating respectively. The trend revealed in this figure shows the FULL treatment to be the most preferred, followed in order by NEW, DFL, and then DFS or RND. We added two 9 point scales to measure the subject's perception of audio and video ("1" = poor audio and poor video). The mean subjective ratings of audio and video are presented in Figure 6.14.

The subjects were directly asked how well they felt the video skim did in communicating the essence of a longer video segment. This question was only asked following the viewing of one of the skim treatments, and the mean results from the 9 point scale ("1" = "inadequately") are shown in Figure 6.15. The subjects were also asked how well they felt the video treatment informed them for answering the text and image questions and these mean results ("1" = "poorly informed") are shown in the figure as well.

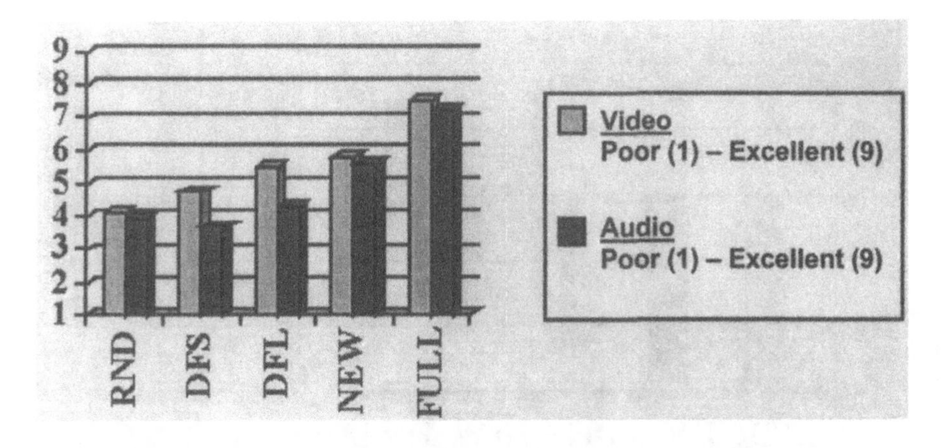

Figure 6.14 Mean scores for subjective ratings on audio/video.

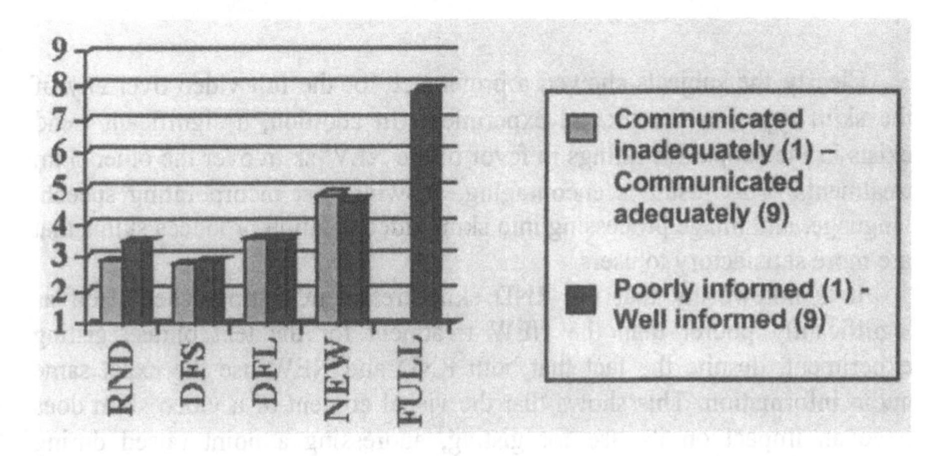

Figure 6.15 Mean scores for two other subjective ratings.

Analysis of variance for the Latin Square Design for all of these seven scales show highly significant ($p < 0.01$) differences in mean subjective ratings among the five video treatments. Testing the means with SNK reveals that FULL's mean is significantly ($p = 0.05$) different from all other treatment means, and that for 6 of 7 cases, NEW's mean is significantly different from the other skim treatment means. For the seventh case ("poor-excellent video") the NEW's mean is still the greatest of the skim treatment means and significantly different from all but the DFL treatment mean.

The subjects' open-ended typed comments support these results as well. An informal and subjective classification of the 59 open-ended comments into a favorable or critical opinion produced the distribution shown in Figure 6.16.

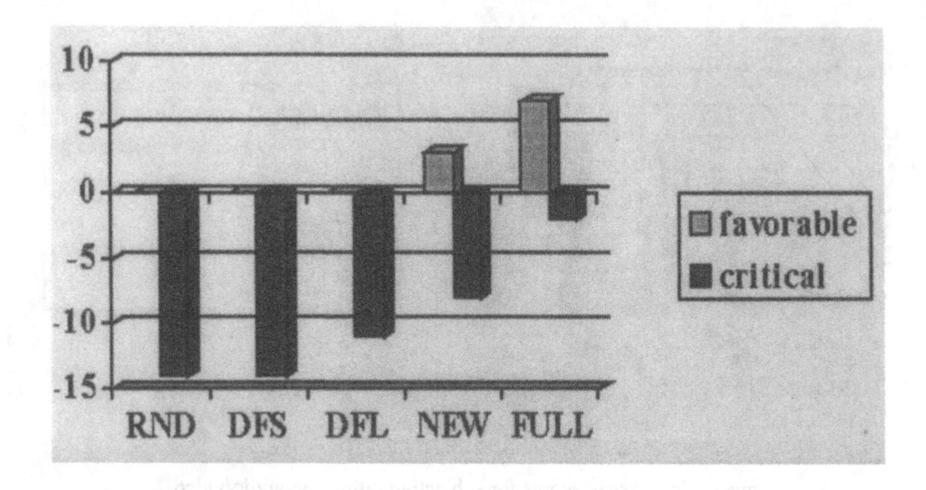

Figure 6.16 Count of open-ended comments by treatment.

Clearly the subjects showed a preference for the full video over any of the skim types in the second experiment. In addition, a significant trend exists in the subjective ratings in favor of the NEW skim over the other skim treatments. This result is encouraging, showing that incorporating speech, language, and image processing into skim video creation produces skims that are more satisfactory to users.

It is noteworthy that the RND skim treatment distinguished itself as significantly poorer than the NEW treatment for the text phrase gisting experiment, despite the fact that both RND and NEW use the exact same audio information. This shows that the visual content of a video skim does have an impact on its use for gisting, addressing a point raised during summer pilot studies.

The DFS and DFL skim treatments did not particularly distinguish themselves from one another, leaving open the question of the proper component size for video skims. The larger component size, when used with signal-power audio segmentation, produced the NEW skim that did distinguish itself from the other skims. If the larger component size is used only for subsampling, however, it yields no clear objective or subjective advantage over short component size skims, such as DFS. In fact, both DFS and DFL often rated similarly to RND, indicating perhaps that any mechanistically subsampled skim, regardless of granularity, may not do notably well.

While very early Informedia skim studies found no significant differences between a subsampled skim and a "best" audio and video skim, this second study uncovered numerous statistically significant differences

[Christel 1998]. The primary reasons for this new finding can be traced to the following characteristics of the audio data in the skim:

- Skim audio is less choppy due to setting phrase boundaries with audio signal-processing rather than noun-phrase detection.
- Synchronization with visuals from the video is better preserved.
- Skim component average size has increased from three seconds to five.

Although the integrated audio and video skim (NEW) established itself as the best design under study, considerable room for improvement remains. It received mediocre scores on most of the subjective questions, and its improvement over the other skims may reflect more on their relatively poor evaluations than on its own strengths. The integrated skim did distinguish itself from the video-reversed skim (RND) for the image recognition and text-phrase gisting tasks, but not from the short evenly sampled skim (DFS) or long evenly sampled skim (DFL). The integrated skim under study achieved smoother audio transitions but still suffered abrupt visual changes between image components. Transitions between video segments should also be smoothed — through dissolves, fades, or other effects — when they are concatenated to form a better skim.

6.5 Poster Frame Study

A separate poster frame study was conducted to compare static and dynamic summaries [Christel 1997]. The experiment tested methods for displaying a result set of video segments matching a given query: conventional text, naively chosen poster frames, and query-based poster frames. A label was used to represent the video segments prior to library exploration. Users were offered one of the following forms of labels:

Treatment 1 (T1) Poster Frames - These are query-based representative images selected using the image understanding techniques described in chapters 3 and 4.

Treatment 2 (T2) First Shot Frame - A first shot frame was selected by choosing the first image of a segment. These frames often contàin credits or blank frames.

Treatment 3 (T3) Text Lists - Text lists are provided in the transcript of these segments. For this test, we extract the initial sentence from a manually created abstract of the transcript.

Each subject received the same on-line multimedia tutorial that discussed the use of the digital library system to accomplish fact-finding tasks. Subjects answered on-line questionnaires concerning their backgrounds. They subsequently received instructions on the screen informing them that they would be presented with three sets of four questions each, which should be answered as quickly yet accurately as possible.

The text on the screen introduced the topic, and then a question was shown. A set of 12 labels from one treatment was displayed. The subject browsed these labels until satisfied that a particular segment was the answer to the given question. Subjects using the query-based poster frames were able to complete the task in the shortest time and with the highest accuracy. Table 6.1 lists the results of this study.

Table 6.1 Results of Poster-Frame User Study

Test Parameter	T1 - Query Based Poster Frames	T2 - First Shot Poster Frames	T3 - Text List
Mean Question Score (Maximum 400)	353.33	346.67	315.00
Mean Time to Complete Question Set (seconds)	590.70	952.00	921.90

6.6 Conclusions

This chapter has described various user studies and testing methods for analyzing the output of a video characterization system. The initial study was conducted with early versions of the video skims. A subsequent study was used to improve and analyze the visual and audio properties of the skim. The techniques and methodology provides an instructional guide for conducting similar studies.

Subjective measures for video analysis provide a standardized method for assessing quality to a video summarization or characterization system. Similar to the studies in language and video quality, the user study will likely remain a prominent component for the study of content-based video analysis.

6.7 Bibliography

Arons, B. "SpeechSkimmer: Interactively Skimming Recorded Speech," *Proceedings of ACM Symposium on User Interface Software and Technology'93*, Atlanta, GA, Nov. 3-5, 1993, pp. 187-196.

Christel, M.G., and Pendyala, K. "Informedia Goes to School: Early Findings from the Digital Video Library Project". *D-Lib Magazine*, September, 1996, http://www.dlib.org/dlib/september96/informedia/09christel.html.

Christel, M.G., Winkler D.B., and, Taylor C.R. "Improving Access to a Digital Video Library". *Human-Computer Interaction INTERACT '97: IFIP TC13 International Conference on Human-Computer Interaction*, 14th-18th July 1997, Sydney, Australia, S. Howard, J. Hammond & G. Lindgaard, Eds. London: Chapman & Hall, 1997, pp. 524-531.

Christel, M.G., Smith, M.A., Taylor, C.R, and Winkler D.B., "Evolving Video Skims into Useful Multimedia Abstractions". *Proceedings of the CHI '98 Conference on Human Factors in Computing Systems*, C. Karat, A. Lund, J. Coutaz and J. Karat, Eds. (Los Angeles, CA, April 1998), pp. 171 - 178.

Degen, L., Mander, R., and Salomon, G. Working with Audio: Integrating Personal Tape Recorders and Desktop Computers. *Proceedings of the ACM CHI'92 Conference on Human Factors in Computing Systems.* (Monterey, CA, May 1992), 413-418.

Hampapur, A., Jain, R., and Weymouth, T. "Production Model Based Digital Video Segmentation". Multimedia Tools and Applications, 1 (March 1995), 9-46.

Mauldin, M. "Information Retrieval by Text Skimming," PhD Thesis, Carnegie Mellon University. August 1989. Revised edition published as "Conceptual Information Retrieval: A Case Study in Adaptive Partial Parsing, Kluwer Press, September 1991.

Mills, M., Cohen, J., and Wong, Y.Y. A Magnifier Tool for Video Data. *Proceedings of the ACM CHI'92 Conference on Human Factors in Computing Systems.* (Monterey, CA, May 1992), 93-98.

Pfeiffer, S. Lienhart, R., Fischer, S., Effelsberg, W., "Abstracting Digital Movies Automatically," *Journal of Visual Communication and Image Representation,* Vol. 7, No. 4, pp. 345-353, December 1996.

Pinson, M., and Wolf, S., "Comparing subjective video quality testing methodologies." *Visual Communications and Image Processing (VCIP 2003),* Lugano, Switzerland, July 8-11, 2003.

QUIS 5.5b. University of Maryland at College Park, 1994. Available through http://www.lap.umd.edu/ QUISFolder/quisHome.html.

Stevens, S. "Next Generation Network and Operating System Requirements for Continuous Time Media". Network and Operating System Support for

Digital Audio and Video, R. Herrtwich, Ed. 1992, Springer-Verlag; New York.

Taniguchi, Y., Akutsu, A., Tonomura, Y., and Hamada, H. An Intuitive and Efficient Access Interface to Real-Time Incoming Video Based on Automatic Indexing. *Proceedings of the ACM International Multimedia Conference.* (San Francisco, CA, November 1995), 25-33.

Wactlar, H.D., Kanade, T., Smith, M.A., and Stevens, S.M. "Intelligent Access to Digital Video: Informedia Project," *IEEE Computer*, 29, May 1996, 46-52.

Yeung, M., Yeo, B., Wolf, W., and Liu, B. "Video Browsing Using Clustering and Scene Transitions on Compressed Sequences". *Proceedings IS&T/SPIE Multimedia Computing and Networking*, February, 1995.

Zhang, H.J., Smoliar, S., and Wu, J.H., "Content-Based Video Browsing Tools," *Multimedia Computing and Networks,* . Multimedia Systems, 2, 6 (1995), 256-266.

Chapter 7

Conclusions

The video skim is an effective tool for video summarization and creation of short preview video that contains content specific audio and image information. Compaction rate as high as 35:1 is achieved without apparent loss in content. Automated semantic solutions to video summarization are plausible with a relatively small number of hierarchical rules. These rules are derived from production standards that evolve as new methods for video production are developed.

Multimodal Features

The utility of integrated multimodal analysis is evident through video characterization, and applications to video summarization. Multimodal integration is a better method for video understanding than individual image, audio or language understanding. These separate techniques, when combined through a subjective rule hierarchy for video prioritization, improve video characterization.

Individual features provide useful characterization for certain applications. Image understanding techniques, which include shot change detection, image similarity, texture analysis, camera and object motion analysis, video caption detection and recognition, and human face detection, are useful for applications specific to images in video. Methods in image analysis with compressed video facilitate processing of large data sets in a short time.

Language and audio understanding improve semantic characterization in video. Language understanding techniques enhance keyword detection and prioritization. Audio analysis provides keyphrase boundaries and discrimination between useful human speech and other sound effects.

Video Skimming

The skim is a summarization method based on video content. Audio and image regions selected from the original video and prioritized based on integrated multimodal video properties form a shorter video that yet conveys the video content. Several summarization techniques were introduced as alternative skim representations in chapter 4. System prototypes and visualizations were presented for testing and evaluating the skim with several hours of digital video in chapter 5. These systems are interactive to allow for user control, variable rate compaction, and use in a digital video library.

User studies are necessary when evaluating subjective video representation systems. The utility of the skim for browsing and previewing digital video was proven through a series of user-studies and interface experiments in chapter 6.

Custom Systems

The general problem of content analysis goes beyond the methods for video understanding described in this book. When the media or video of interest is limited to a specific domain, custom applications may benefit from higher quality feature analysis and visualization interfaces. Custom video summary systems are viable applications in sports video, news highlights, movie trailers and many venues, as described in section 2.4.

Future Characterization

The video skim and the visualizations described in chapters 4 and 5 are scalable and modular. The subjective quality of visualization will increase as video content analysis improves. The feature analysis described in chapter 3 provides an introduction to the current state-of-the-art in automated methods for video characterization. As these techniques improve, so shall the quality of video summaries and other visualization applications that use multimodal features. Figure 7.1 is a depiction of a small set of working and experimental characterization features.

Shot Cuts			
Camera	Static	Static	Zoom
Objects	Adult Female (1)	Animal	Male Adult (2)
Action	Head Motion	Left - Walking	[None]
Captions	[None]	Yellowstone	[None]
Scenery	Indoor	Outdoor	Indoor
Gesture	Neutral	[None]	Smile

Figure 7.1 Characterization Research Activities.

Conclusions

The emergence of high volume video libraries has shown a clear need for content-based video characterization and summarization technology. Most video understanding systems place emphasis on image understanding technology, but audio and language understanding provides greater potential for semantic characterization and summarization.

True video understanding is a difficult problem, but the methods described in this book illustrate the power of integrated audio, language, and image information for characterization in video retrieval, browsing and summarization applications.

Index